Technology for the Common Good

Technology for the Common Good

Michael Shuman

&

Julia Sweig

Editors

Institute for Policy Studies
Washington

The Institute for Policy Studies is a non-partisan research institute. The views expressed in these essays are solely those of the authors.

Library of Congress Catalog Card Number: 93-078140

ISBN 0-89758-047-8

Manufactured in the United States of America

Copies of this book are available from
The Institute for Policy Studies
1601 Connecticut Avenue, NW
Washington, DC 20009
(202) 234-9382

Contents

Preface

Nineteen sixty four was a watershed year for the United States. Camelot came crashing to a halt when an assassin's bullets struck down the nation's youngest president. The Civil Rights Movement was in full swing, the War on Poverty had just begun, and another war in Southeast Asia was looming. It was a time of enormous despair—and enormous opportunity.

In March of that year a committee of eight thinker-activists came together to draft a visionary statement called "The Triple Revolution." The document ultimately bore the signatures of thirty-five noted opinion-leaders such as Robert Heilbroner, Irving Howe, Linus Pauling, Gerard Piel, Bayard Rustin, and Norman Thomas. It warned of incipient economic, racial, and foreign-policy crises and presented a long list of remedial policies. In part because of its bold recommendation that the country establish a guaranteed annual income, "The Triple Revolution" became an overnight sensation. It was front-page news in *The New York Times* and sparked a lively debate on op-ed pages across the country.

"The Triple Revolution" described three interrelated developments:

- *"A Cybernation Revolution,"* in which computers and machines were creating "a system of almost unlimited productive capacity which requires progressively less human labor."
- *"A Weaponry Revolution,"* in which nuclear weapons capable of incinerating civilization had "eliminated war as a method for resolving international conflicts."
- And *" A Human Rights Revolution,"* in which people everywhere, particularly people of color, were demanding full and equal rights.

"The Triple Revolution" predicted that technological development without national direction, public jobs programs, or effective redistribution policies would result in greater inequality and relegate people of color to a permanent underclass. It called for a guaranteed annual income above the poverty line instead of a "patchwork of welfare measures," as well as for a far-reaching program of economic planning and public works. And it eschewed piecemeal arms control agreements and stressed that only comprehensive disarmament could free up scarce national resources for needed housing, transportation, and energy production.

Today, nearly thirty years after "The Triple Revolution" was first articulated, all three revolutions are critically important for public policy.

The details of course have changed, but the basic themes haven't. Cybernation has occurred on a global scale and resulted in greater income inequality, in a shift of jobs from production into service sectors and information industries, and in massive environmental problems that threaten the survival of the biosphere. War has not become obsolete — in 1987 there were more conflicts raging than at any other time in recorded history — but the winding down of the Cold War has brought the need for disarmament sharply into focus. And even though civil rights have become firmly institutionalized in federal and state laws, women, children, gays and lesbians, immigrants, and the disabled are still struggling for equal protection.

To examine the first of these issues through the historical prism of "The Triple Revolution," the Institute for Policy Studies decided to hold an invitational conference, "Technology for the Common Good." Once again, a group of progressive thinker-activists came together (on October 4-5, 1991) to craft a common agenda, only this time there were three dozen of us. We discussed five interrelated questions:

- What should politically "progressive" criteria for evaluating technology look like?
- What should be the nation's research and development priorities?
- How can the government better regulate technology?
- How can public participation on technology questions be improved?
- What kind of strategy should grassroots groups undertake to implement this agenda?

We focused on technology for two reasons. First, this was the subject which seemed to lie at the heart of "The Triple Revolution." Second, the subject remains an extremely timely one. Technology has profound implications for every facet of life yet receives little attention either from the public policy community or from activists. Over the past generation powerful citizen movements have arisen to push for banning the bomb or enacting civil rights, but there is not yet a movement for more democratic, equitable, and sustainable technologies.

Another catalyst for revisiting "The Triple Revolution" was to honor its organizer, W. H. Ferry. Ferry's persistence in asking tough questions about technology, year after year, convinced us to launch this endeavor. His seminal thinking and writing laid the foundation for the conference and this book.

What follows are a series of essays that give a flavor of the issues raised at the conference. The first piece is an overview entitled "Responsible Technology for a New World Order," which was drafted to synthesize the ideas of all participants at the conference. The remaining essays focus on topics of special importance or interest to progressives:

- Joel Yudken critiques the policy of the Bush Administration to invest only in technologies with military usefulness and suggests how the Clinton Administration should refocus national research and development expenditures on critical social and economic needs.
- Richard Sclove argues that if Americans wish to preserve their democratic rights, they must deliberately choose technologies that are more compatible with democracy, through more democratic means.
- Charley Richardson analyzes the pernicious influence of Taylorist design principles on the workplace and the benefits of giving workers greater say over the technologies they use and produce.
- Sanford Lewis examines the dangers of technologies that rely on toxic chemicals and recommends a range of public policies for ameliorating these dangers.
- Chellis Glendinning looks at how the victims of technology can fight back.

We also added appendices that readers might find useful, including the text and signatories of the original Triple Revolution statement and a list of the participants in our own conference. We are grateful to a number of funders who helped support this project, including the Aaron Diamond Foundation (directed by Vincent McGee), the New Land Foundation (Hal Harvey), and Rockefeller Family Associates (Wade Greene). We also wish to thank the trustees, fellows, and staff of the Institute for Policy Studies for supporting this endeavor long past its original schedule.

Unauthorized reproduction is *encouraged* (though extra copies of the entire book can be ordered using the form in the back of the book). We urge readers to share this book, particularly the overview statement on "Responsible Technology for a New World Order," with politicians, policymakers, scientists, engineers, lawyers, economist, journalists, and activists everywhere. As Albert Einstein once said, to the village square we must take the profound questions raised by technology, and from there must come America's voice.

Michael Shuman & Julia Sweig, April 1993

Introduction

Responsible Technology for a New World Order

For nearly half a century the United States marshaled its people, resources, and know-how to win the Cold War with the Soviet Union, but victory did not come without cost. One significant casualty was the nation's policy toward technology, which was wrongly based on militarization, megaprojects, and marketism:

Militarization—Literally hundreds of billions of dollars were poured into universities and corporations to design, test, and manufacture a dazzling array of missiles, bombers, tanks, submarines, and other weapons.

Megaprojects—What little federal money did not go for weapons research and development (R&D) went to large, flashy, but ultimately worthless boondoggles like the Clinch River Breeder Reactor, the Synthetic Fuels Corporation, and the Hubble Telescope.

Marketism—Beyond military R&D and civilian megaprojects, the government wrongly assumed that the free market would provide all the technologies needed for our well-being.

Calling these three M's a technology policy probably gives national leaders too much credit. As Frank Press, President of the National Academy of Science, noted in 1988: "It is astounding but true that nowhere in the federal budget-making process is there an evaluation of the complete federal budget for science and technology and its overall rationale in terms of national goals."

Today, we call for an entirely new technology policy for the United States based not on militarism, megaprojects, or marketism, but on social responsibility. We call for the government to promote technology that is responsible to the basic economic, social, and political needs of Americans, as well as to those of people throughout the world. And we urge local, national, and international leaders to develop and invest in responsible technology.

Responsible technology begins with the recognition that all technology is the result of human choice. Technology is a human activity and like other human activities it can have both good and bad intentions and consequences, it can affect different people or different interests in unequal ways, and it can be supported or regulated as much as the national government or communities wish. Nothing about technology is preordained, unless we choose to do nothing about it.

A commitment to responsible technology requires us to embrace four basic principles:

- Technology should be judged on the basis of its impact on the common good of all people, in particular whether it addresses the urgent crises faced by the nation and the world today.
- International, national, and local plans should be developed that encourage investment in technology that benefits the common good.
- The government has a responsibility to outlaw or regulate technology that imperils the common good.
- And in a democratic society each of these tasks—judging, financing, and regulating technology—should be undertaken with the widest possible level of grassroots participation.

Five Crises in America

Today there is widespread belief that scientists should undertake every interesting experiment and that business should produce every profitable gadget. All technology is viewed as progress. In fact, however, much of what masquerades as "progress" has moved the human race backwards.

Whatever the value of the three M's in the past, they now comprise a fundamentally inadequate national policy. They fail to address, and may well exacerbate, five crises facing the country:

The Weapons Crisis

Militarization has enabled the United States to assemble the most powerful weapons arsenal in the world, yet the country is not more secure. Our weapons buildup spurred the Soviet Union to acquire a fearsome arsenal of nuclear weapons, which is now spread among three unstable former republics. It fueled a massive international arms trade, which has placed the technology for nuclear, chemical, biological, and advanced conventional weapons in the hands of dozens of countries and perhaps even terrorists. And it saddled the country with thousands of contaminated weapons-production facilities and radioactive waste dumps.

The Economics Crisis

Militarization also has created a crisis in the American economy. One by one, the nations we spent billions to defend during the Cold War are surpassing us in commercial power. Because the spread of nuclear weapons has rendered war obsolete, competition among great powers will no longer be played out on the battlefield but instead in the global marketplace. And in that competition we're losing badly. While we've been investing two thirds of our national R&D monies on better weapons, Japan and Germany have been investing in better cars, steel, VCRs, and toasters. And while we've slashed domestic programs to pay for ambitious foreign policies, our competitors have spent considerable sums to ensure that their citizens are healthy, housed, fed, and well educated. To continue spending federal R&D money on military gadgets instead of commercial technology will ensure the continued deterioration of our economy and society.

The Ecological Crisis

Competitiveness, however, cannot be the sole basis for a new technology policy. The current level of production in the world, fostered largely by our success in spreading the gospel of marketism, has already punched gigantic holes in the earth's ozone layer, denuded vast expanses of rain-forest, killed off thousands of species of plants and animals, and unleashed the specter of global warming. If the world continues economic business as usual, a simple walk outdoors will be a serious cancer risk, ocean levels will rise and inundate coastal areas, vital farming regions will become deserts, and entire ecosystems will collapse. Efforts to develop technology for more consumption and production may well erode the life support systems necessary for human survival.

The Work Crisis

The standard prescriptions for competitiveness also do not address, and may well worsen, growing problems in the workplace. The automation and computerization of factories and offices has meant unemployment for millions of Americans. For two-thirds of the American workforce, wages have dropped twelve percent over the past fifteen years. The globalization of the world's economy, another result of unfettered marketism, means that every country must now compete for jobs provided by multinational corporations that are accountable to no single country. This competition to lure corporations is inducing communities to cut wages, bust unions, weaken worker safety requirements, and dilute environmental protection measures. "Competitiveness" has become the excuse for lowering the standard of life for people everywhere. The magic of the free market can no longer assure full or satisfying employment for working people in America.

The Democracy Crisis

The collapse of state socialism has not led to the triumph of democracy, especially in technology policy. Even though decisions concerning technology affect almost everyone, our adherence to the three M's has meant that these decisions are made largely in Pentagon offices, government bureaucracies, or corporate boardrooms, thousands of miles away from the people affected by them. We are told that matters of technology are too complex for lay citizens to understand, that we should entrust these choices to scientists, engineers, and economists. We have done so, in part because an education crisis has robbed many of our fellow citizens of the capacity to think critically or to act as informed citizens. A country that once prided itself as the world's model democracy now finds its citizenry increasingly narcotized by television, deprived of public spaces for organizing, and turned off by electoral politics.

Together, these five crises are eating away at the fabric of American life. If they continue unaddressed, they could culminate in war, economic depression, ecological calamity, widespread unemployment, and authoritarian rule. These crises already can be seen in the form of social breakdown, as Americans cope with unprecedented levels of stress, divorce, sexual abuse, and crime. It's little wonder that so many of our young people feel disaffected, alienated, and hopeless (suicide is now the second leading cause of death among those between the ages of 16 and 24).

But we have the power to slow the onset of these crises, and perhaps the power to solve them, if we are willing to move in a fundamentally new direction: We must take control of technology. We must stop the development, production, and dissemination of irresponsible technology and begin the development, production, and dissemination of more responsible technology. While responsible technology is not a panacea, it is a necessary and important part of any larger set of solutions to the five crises. Technology has become too inextricably interwoven with every aspect of society to be overlooked any longer.

Criteria for Responsible Technology

We call upon citizens everywhere—as voters and politicians, as consumers and producers, as workers and industrialists—to scrutinize technology more critically. Technology must be evaluated by all its impacts: intentional and unintentional, direct and indirect, immediate and long-term. Our willingness to promote or allow a given kind of technology should hinge on whether these impacts advance other societal goals. Responsible technology, in our view, should meet the following six criteria:

Peace—To end the legacies of militarization—namely the weapons crisis and the economics crisis—there should be a substantial reduction in government investment in military technology. Now that the Cold War is over, the moment is ripe to stop the development and production of all nuclear, chemical, and biological weapons—weapons which have no other purpose except to murder human beings *en masse*. In addition, the development and production of weapons capable of offensive military uses, such as battleships, bombers, long-range missiles, and tanks, should cease. The only weapons necessary for national security are those which serve purely defensive purposes.

Common Welfare—Ending the crisis afflicting our economy requires not only reduced expenditure on military technology but also increased expenditure on technologies that can solve our underlying economic problems. We endorse the views of many leading economists that renewed national investment is needed in education and the social safety net. Better education means a better work force, wiser consumers, and more politically engaged citizens. Because it is impossible for people to learn, let alone work or participate in politics, if they are hungry, sick, or homeless, it is essential to provide every American with decent food, health care, and housing. But economic growth must be built on sustainable industries.

Sustainability—We must rebuild our economy around technologies that do not worsen, and can solve, the world's ecological crisis. We should favor technologies that require the fewest inputs of scarce materials, rely primarily on renewable resources, and pollute the least (or preferably not at all). Many of the economic activities that nations now regard as essential will have to be dramatically changed or abandoned to meet this criterion. But this criterion also opens up new business opportunities for recycling, environmental restoration, and renewable energy production.

Jobs—To meet the work crisis, we must choose technologies that improve the quantity and quality of jobs. No American should be denied a decent job. Public employment programs should produce goods and services that help communities resolve conflicts, improve their quality of life, clean up their environment, and enrich local democracy. Moreover, we should encourage every workplace to adopt technologies and production methods that allow individuals to perform work that is creative, non-repetitive, safe, and satisfying. None of this will be possible unless workers and trade unions are given more power to participate in decisions over what products to produce and how to produce them.

Strong Democracy—Public participation over decisions governing technology is necessary if we are to overcome the crisis in democracy. Participation means much more than showing up at the polls once a year. It also means working with policymakers on an ongoing basis and *becoming* a policymaker from time to time. Strong democracy requires both new institutions and new technologies. New institutions are needed to increase the public's technological literacy and to oversee technological decision-making. And new technologies such as interactive television are needed to link people more effectively with one another and with policymakers.

Self-Reliance—A key to solving today's crises is to empower citizens to act at the local level, without needing to rely on outsiders. A truly self-reliant community poses no military threat to others. A community where people know their neighbors and can see how their tax dollars are being used can muster the political will to protect the welfare of every resident. A self-reliant community, by definition, would not depend on scarce resources from outside the community and would not dump pollution or waste on its neighbors. Finally, a self-reliant community gives people enough power so that they want to participate in democratic decisionmaking.

We recognize that all these criteria are subject to varying interpretations and sometimes conflict with one another. It is important, therefore, that democratic processes be designed to ensure that these principles are applied to technology decisions with the fullest public participation possible.

National Investment in Responsible Technology

The U.S. government has demonstrated through its support of militarization and megaprojects that it can play an important role in setting national priorities. These priorities of the past have benefitted defense contractors and large scientific establishments but not the general public. To serve the common good, the nation's technology policy must be changed fundamentally—not to a policy that emphasizes competitiveness but to one that emphasizes America's real needs.

Misguided National Priorities

For the past decade the U.S. government spent roughly two-thirds of its R&D monies on military projects (about 40 percent of all R&D monies nationwide), while Germany and Japan spent most of their national R&D budgets on nonmilitary projects. A quarter of our young scientists and engineers worked on defense projects, while the Germans and Japanese harnessed their brightest minds to build better cars, computers, and airplanes. The predictable result is that the United States has become the world's best producer of weapons, while Europe and Japan have become the best producers of commercial goods.

Until we demilitarize our technology priorities, we will continue to fall behind other nations in nonmilitary productivity. A growing body of evidence suggests that military spending produces fewer jobs and fewer competitive products than does nonmilitary spending. Moreover, commercial "spin-offs" from military R&D are becoming more difficult for several reasons: military investment decisions are rarely made with spin-offs in mind; the specifications for complex weapons systems are becoming less applicable to commercial products; and the secrecy and export controls surrounding military innovations keep them out of civilian markets.

Those within government who want to increase the number of commercial spin-offs from defense R&D have tried to strengthen the Defense Advanced Research Projects Agency (DARPA). But whenever DARPA wandered into projects to produce "dual-use" technologies, such as high-definition television, X-ray lithography, high temperature superconductors, or new semiconductors, the Bush Administration squelched it. Congress and the Clinton Administration have been more enthusiastic about DARPA's efforts to seed dual-use technologies, but DARPA is an inefficient tool to strengthen America's nonmilitary industries. If DARPA-funded researchers spend half (or more) of their time developing military technology while researchers in Europe and Japan spend nearly all their time developing commercial technology, we will still be losing ground to our competitors.

Nearly all the geopolitical reasons that the Cold Warriors once gave for pouring so much government money into military R&D have vanished. The Soviet threat, which provided the rationale for most military R&D, is now gone and splintered into fifteen new countries. For the first time since World War II, the United States has an opportunity to withdraw its forces from Europe, to enter into comprehensive disarmament agreements, and to transfer the burdens of global peacekeeping to the United Nations. The Bush Administration reluctantly agreed to trim the defense budget by 25 percent over the next five years, and the Clinton Administration plans to trim a little more, but a wide range of experts now say that a cut of fifty percent is possible. An additional $150 billion per year would enable the United States to eliminate the federal budget deficit and expand a variety of economic and social programs, including those related to responsible technology.

The other misguided priority for federal technology policy has been "big science." A growing fraction of the federal government's nonmilitary R&D budget is being earmarked to six megaprojects—the Strategic Defense Initiative, the Space Station, the Moon/Mars Initiative, the Superconducting Super Collider, the National Aerospace Plan, and the Human Genome Project. These projects are likely to generate few social benefits and are funded at the expense of thousands of smaller endeavors that could address urgent social needs. Many scientists privately concede that the government's priorities are irrational, but the federal system of competitive grants has silenced all but the most courageous critics.

The Folly of Competitiveness

Many are eager for the national government to reorganize its spending priorities under the rubric of "competitiveness." To be sure, there is much to commend in policies that strengthen the relative attractiveness of American products to buyers here and abroad. But the term is vague and has been used to support all kinds of economic deal-making, regardless of the impact on the common good.

Simply boosting sales cannot remedy many of the dangers that lie ahead. If competitiveness means selling more armaments to aggressive nations, it will exacerbate the weapons crisis. If competitiveness means weakening environmental regulations or lowering the wages of American workers, it will worsen the ecological and work crises. If competitiveness means robbing communities of their ability to regulate business and entrusting those powers to an omnipotent General Agreement on Tariffs and Trade (GATT), then it will stoke the democracy crisis. Competition can only be healthy if the game is fair, and right now international rules protecting workers, the environment, consumers, and communities are practically nonexistent.

There are certainly worthy national investments that could make American goods and services more competitive. As noted earlier, government spending on the education and welfare of all Americans and on sustainable industry might be especially fruitful. But a broader vision of national priorities is needed.

A Fund for Responsible Technology

We call on the national government to create a special fund to assist with the research, development, and production of responsible technology. There is no shortage of technological projects on which government money could be wisely spent. High on our list would be energy efficiency devices, solar-power and wind-power generators, fuel-efficient automobiles, mass transit systems, recyclable materials, toxic-waste cleanup, and environmental restoration techniques. We also see value in investing in the fields of psychology, sociology, anthropology, and political science to better understand the subtle effects of technology and to develop improved democratic tools for debating social decisions concerning technology.

Consistent with our criteria for responsible technology, the national government should set up a competitive grant process with the following guidelines:

- To promote peace, no funding would be given to technologies with direct or indirect military uses.
- To promote the common good, different grant categories for technology development would be created to correspond with specially targeted needs in education, health care, housing, and environmental protection.
- To promote sustainability, every applicant would have to demonstrate that the technology being developed will help society live within its ecological means.
- To promote work, every applicant would have to demonstrate that the technology could improve the quantity and quality of the nation's jobs.
- To promote strong democracy, the decisionmaking body should include representatives from business, labor, universities, women's groups, minorities, and public interest organizations.
- And to promote self-reliance, at least half the fund should be administered by citizen boards set up by city or county governments.

Resources to support this fund would come from cutting the most unproductive projects upon which the government now lavishes military R&D monies. New money would also become available by scaling back big-science projects or undertaking them through multinational consortia.

Another way the government could promote responsible technology is through tax incentives. For example, corporate expenditures to develop technologies that conserve energy or protect the environment could be depreciated more quickly and the gains from these technologies could be taxed at a lower rate. Increasingly, the obstacles to cutting the military budget and big science are local economic interests. Communities and businesses that are dependent on military bases, defense contracts, or federal laboratories legitimately fear economic dislocation, and it's understandable why they react by lobbying, however irrationally, to maintain high levels of Pentagon spending on the military and big science. This underscores why "conversion" legislation is essential to help ease the transition to a demilitarized economy for workers, businesses, and communities.

Regulating Irresponsible Technology

Responsible technology means that not every science experiment that can be done should be done, and not every product that can be developed should be developed. The world would be a far safer place today if forty-five years ago nations had agreed not to develop hydrogen bombs, MIRV'ed missiles, nerve gas agents, and anthrax-filled artillery shells. But the problem goes beyond the weapons laboratories. Many nonmilitary innovations from corporate or university laboratories have had dramatic social, economic, and political "externalities." Civilian nuclear power plants, for example, continue to pose major public hazards through meltdowns, waste disposal, and nuclear weapons proliferation. State-of-the-art communications and computer technologies threaten privacy and civil liberties. The disastrous Bhopal chemical leak, the Valdez oil spill, the Love Canal toxic dump, and countless jumbo jet crashes all are reminders of the hidden costs of technology. And almost every technology has social impacts that we rarely count. Television, for example, drowns out family conversation, teaches children to resolve conflicts with violence, and encourages people of all ages to buy products they never knew they needed.

Many economists, such as those in President Bush's Competitiveness Council, have attacked regulation for making products more costly. But a realistic accounting of all the hidden costs of technology—environmental, social, and political—would show that substantial savings could accrue from better regulation. It's also worth emphasizing that much of what really matters is not the bottom line but the quality of life—the beauty of nature, the survival of whales and birds, the psychological benefits of a society free of racism and sexism — none of which can be counted in dollars and cents. Government regulation, therefore, must be based not just on cost-benefit analysis but on the kind of society we want. Here, the six criteria outlined earlier provide important guidance.

But once we have chosen the goals of regulation, deeper questions remain about means. Today, government regulation of technology generally is too little, too late. Only after the horrors of a technology appear do we begin to think about how to control it. But by that time, entrenched vested interests are already pushing for the expansion of the technology. Public interest groups may coalesce to highlight particular problems, but they are relatively small, fragmented, and underfunded—rarely a match for the formidable battalions of lobbyists pushing for the technology. Moreover, once a technology goes commercial, the market exerts a pull that is difficult to counter. When citizen groups manage to stop a technology after it goes commercial, everyone bemoans how much could have been saved had effective regulation been implemented earlier. We believe that regulation can be invigorated by giving citizens and communities three basic rights: the right to be left alone, the right to know, and the right to inspect and regulate.

The Right to Be Left Alone

One way to regulate technology more effectively is to protect the rights of individuals. Take the right of privacy. More and more communication is taking place over computer networks and portable telephones that can be surreptitiously intercepted. The new "Caller ID" feature on telephones automatically reveals the telephone number of a caller to the receiver. An expanding number of facts about our private lives are being stored by government agencies, banks, credit evaluators, health insurers, junk mail houses, and even video stores. In early 1991 the Lotus Corporation almost released a software package containing detailed information (name, address, age, social security number, marital status, buying habits) on 120 million Americans; only massive public outcry convinced the company to kill the product. National legislation is necessary to protect certain pieces of personal information from being collected or sold without the express permission of the named individuals.

The Right to Know

Knowledge is power. We therefore endorse regulations that would require government agencies and private firms to collect, publish, and disseminate vital pieces of data about their operations. Unlike individuals, whose rights to privacy should be protected, government agencies and corporations exist to serve the people. Here are some examples of the kinds of information that might be required of them:

- To promote peace, agencies and firms might describe each contract or subcontract on a military project.
- To promote the common good, agencies and firms might explain how their products or services are helping to improve the quality of life in the United States or the world.
- To promote sustainability, agencies and firms might document their consumption of land, materials, and energy, and their outputs of pollution, waste, noise, and heat.
- To promote decent jobs, agencies and firms might show how many jobs they are producing for every dollar invested, what kind of jobs these are, with what kinds of wages and what kinds of hazards.

- To promote strong democracy, agencies and firms might examine how their workplaces and products were affecting the ability of citizens to participate in community affairs.
- And to promote self-reliance, agencies and firms might analyze whether they were helping surrounding communities develop independent and vibrant economies (and, conversely, whether they were fleecing communities by shutting down or relocating plants with little notice).

Some of this information is already in the hands of the federal government but lies buried in reports, studies, memos, and letters. One remedy is to strengthen the Freedom of Information Act (FOIA) and to waive copying fees for watchdog organizations. Each agency's filing system should be comprehensively described and publicized for those seeking information. And the statutory exceptions for FOIA requests should be narrowed. For years, for example, the Pentagon hid behind the FOIA exception for "national security" matters to elude public scrutiny of its toxic waste problems. The Departments of Defense and Energy should be required to release all information pertaining to worker and public safety that would not divulge state secrets.

The Right to Inspect and Regulate

Citizens living near government or corporate workplaces should have more than a right to know—they should also have a right to act. They should be able to examine records, to inspect polluting facilities, and to hold hearings on special hazards. And if citizens discover information that differs from what has been publicly reported, they should be able to mount "private attorney general" lawsuits to penalize the agency or firm with criminal and civil sanctions.

Every community should have the right to regulate a plant operating within its boundaries. This would include the right to shut the plant down after careful consultation with workers at the plant and with citizen groups. A community also should be able to require that a company with a plant operating in its jurisdiction add seats to its board of directors for community representatives. And it should be able to demand that a company provide at least 90 days notice before shutting down a plant. There is a need for stronger national and international regulation of multinational corporations. As long as a company can move its operations to unregulated "enterprise zones" overseas, it will be difficult for communities to exercise much control over corporations. It is therefore essential that the United States take a leadership role in creating a set of global rules governing the behavior of companies. These rules would set minimum wages, protect worker rights and other human rights, and spell out basic standards of environmental protection. A new social charter could be inserted into GATT that allows one country to slap punitive tariffs onto any country violating these standards.

Global standards must become floors for further regulation, not ceilings. Countries, states, and communities should be able to set more vigorous standards, provided they are applied on an equal basis to local and foreign goods. Thus, even if a global minimum wage of $1 per hour were set, the United States could retain its minimum wage of $4.35 per hour, and Los Angeles could set a minimum wage of $10 per hour.

Every nation, state, and community also should retain the right to ban any product deemed contrary to the public interest. A key lesson learned by those fighting pesticides, chlorofluorocarbons, asbestos, waste incineration, and various toxic chemicals is that regulatory laws are often broad enough for administrative agencies to subvert their purposes. Because agencies are usually captured by the industries they regulate, regulatory frameworks fail to protect the public adequately. Outright bans, in contrast, cannot be subverted and put pressure on industry to develop safe and environmentally sound alternatives.

We believe that certain broad classes of technology should be prohibited until their safety can be demonstrated. We should treat toxic chemicals, pesticides, radionuclides, and biotechnology like prescription drugs: Ban them until their risks are proven acceptable to workers, consumers, and the environment.

Increasing Public Participation

Finally, each and every American must assume a role in creating a responsible technological order. Too often we have left society's decisions about technology to the "experts." To be sure, experts are needed to sift through piles of data underlying technology decisions and to translate complex facts into terms that lay people can understand. But the right to weigh and evaluate these facts belongs to citizens. Every American should have the opportunity to assess the pros and cons of different technologies, to determine how much government investment they deserve, and to frame appropriate regulations. Popular control over technology needs to be asserted in the community, in the workplace, and in people's personal lives.

Community Democracy

Every state and every community can become a crucible for designing and disseminating responsible technology and for regulating or banning irresponsible technology. We encourage localities to empower committees of citizens to examine the existing technological order, to study alternatives, to invest in promising new technologies, and to suggest regulations. Local governments should implement the recommendations of these committees using all of their conventional powers—education, lobbying, zoning, policing, investing, contracting, purchasing, and litigating. Even when a local government cannot act directly, it can nevertheless encourage citizen action. Thus, for example, while federal law prohibits communities from closing nuclear power plants on the basis of health and safety concerns, communities are not foreclosed from shutting down these plants on economic grounds. Nor are communities precluded from conserving energy (which effectively can put reactors out of business) or punishing reactor manufacturers with citizen boycotts or pension fund divestment.

Community technology planning should be integrated into local economic planning. A city or county should choose technologies that serve the well being of all its inhabitants, that enable it to live within its ecological means, that provide decent jobs to everyone who wishes to work, that can engage its citizens in local decisionmaking, and that help the community become more self-reliant.

It is not enough for a community to use responsible technologies. It also must design or tailor them for local needs. And to promote local self-reliance, these technologies should rely on community resources and talent. Thus, for example, when Chicago began to renegotiate its electricity contract with Commonwealth Edison, it rightly demanded not only that the utility undertake energy efficiency measures but also that these measures involve local materials, local contractors, and local financing.

We encourage communities to undertake creative, democratic experiments to discover new ways to increase citizen participation in technology decisions. Design competitions for technologies could be set up. A limited number of "citizen sabbaticals" might be awarded each year to enable grantees to acquire new skills for exerting community leadership on technology questions. A local computer network could be created, as the city of Santa Monica, California, has done to facilitate joint problem solving by citizens and city staff. Citizen boards could be impaneled, as the cities of Berkeley, California, and Burlington, Vermont, have done to address special technology questions. Grassroots groups could work more closely with municipal designers, just as disabled citizens have done to ensure that sidewalks and public buildings provide wheelchair access. Neighborhood "science shops" could be opened up, as the Netherlands has done, to help people devise technical solutions to special problems.

But for participatory democracy to work at the local level, major political reforms will be necessary too. More public spaces for meeting and organizing are necessary. If disadvantaged groups, such as women and people of color, are to participate fully in decisionmaking panels, they will need financial assistance to take time off work and to hire needed child care. Campaign reform may be necessary to ensure that minorities are protected and that the wealthy do not have disproportionate influence over political processes. Courses on the social impacts of technology may be needed to help educate the public about why participation in technology policy is so important.

Workplace Democracy

We further believe that technological innovation requires a new relationship between workers and managers. The participation of labor on corporate boards, management committees, and design teams can create a more humane and innovative workplace. When the workers at Lucas Aerospace, a defense contractor in the United Kingdom, faced massive layoffs, they organized themselves into design teams and eventually came up with 150 new product ideas (management, unfortunately, was unreceptive). Labor laws in the United States need to be redrawn to resemble those of Sweden or Germany, where management is required to achieve a high level of worker participation.

Another way to democratize corporations is to give citizens and communities greater power in corporate decisionmaking. Once firms exceed a certain size, they should be required to provide a certain number of board seats to community representatives.

Personal Empowerment

Ultimately, control over technology must begin at home. Each of us relies on technology for almost everything we do—washing, cooking, driving, communicating, working. We can influence others by creating the kind of technological order we want for ourselves. If we want more efficient transportation, for example, we should ride our bikes to work, use mass transit, and purchase cars with excellent gas mileage.

Besides setting a good example, we can each exert influence in other simple ways. We can educate ourselves, our family, and friends about what kinds of technology are responsible and what kinds are not. We can teach ourselves and each other about how to become more effective citizens. We can vote for politicians who promise to create a responsible technology order, and we can assist them once they take office. We can refuse to buy goods, services, or securities from any corporation that produces or uses irresponsible technologies.

But perhaps the most radical change necessary is to make a political commitment. The entire way that we conduct politics must be overhauled. We must challenge politicians to devise new ways to engage citizens, and more of us must be prepared to run against those politicians who refuse to open up the political system.

The Task Ahead

It will not be easy to transform the current technological order into a more responsible one. New efforts to judge, invest in, and regulate technology properly will take time, energy, and money, especially if they are to involve widespread public participation. But the costs of not pursuing this agenda are even greater. The continued reign of irresponsible technology will mean more weapons proliferation, more global poverty, more ecological destruction, more unemployment, and more personal and community disempowerment.

In the early 1960s economist Robert Heilbroner wrote: "[T]he coming generation will be the last generation able to seize control over technology before technology has irreversibly seized control over it. A generation is not much time, but it is some time." A generation has since come and gone, and the damage wreaked by irresponsible technology grows greater every day. We call for compassionate people everywhere to seize control of science and technology today—we will not have another chance.

Chapter 1

Beyond "Dual-Use" Policy: A National Needs Investment Strategy for Science and Technology

by Joel Yudken

With the end of the Cold War, Americans confront three fundamental challenges with profound implications for their security, economic well-being, and quality of life. First, the United States must restructure its international and military commitments, shrink its massive military-industrial apparatus, and redirect freed economic and human resources to pressing domestic problems. Second, it must revitalize its sagging economic base and bolster its industries in an increasingly competitive global marketplace. Finally, it must respond to an emerging environmental crisis that threatens life on the planet. How can the nation harness its formidable science and technology (S&T) resources to meet these challenges?

Government and business leaders preoccupied with the decline in U.S. "competitiveness" agree that strengthening the nation's S&T base is essential to improve its industrial performance. There are strong differences of opinion, however, over how this should be done, particularly over the role of the government in supporting commercial technology. A growing chorus of high-tech managers, academic experts, and public officials are calling for greater government assistance for the nation's faltering manufacturing industries, especially in the high-tech sector.[1] This group wants increased federal investment in commercially relevant technologies, more trade protection, greater tax credits for research and development (R&D), relaxed anti-trust laws, reduced regulation, broader intellectual property rights, and new patent laws. Many also are concerned that the lion's share of the nation's R&D investment is being devoted to military projects, while civilian technology remains underfunded. Typifying this group is Brookings Institution fellow John D. Steinbrunner, who contends that with the Cold War over, we need to disconnect national technical investment "from the substantial reliance on national security that it has had for four decades and direct it to economic performance."[2]

A number of proponents for a stronger federal hand in commercial technology have lined up behind the notion of "dual-use" technology policy, which emphasizes government support for R&D on "critical technologies" deemed essential for both civilian and military needs. However, lacking sufficient support from civilian R&D agencies—and given U.S. industry's historical inability to invest in long-term R&D—high-tech proponents have relied on the Department of Defense (DoD), especially the Pentagon's lead R&D agency, the Defense Advanced Research Projects Agency (DARPA), to bankroll most of the dual-use efforts undertaken in the United States to date.

Interestingly, powerful officials from the Bush Administration who were ideologically opposed to any form of "industrial policy" came out against these proposals, including initiatives funded by DoD and DARPA. These critics included former White House Chief of Staff John Sununu, the President's chief economic adviser Michael Boskin, and Office of Management and Budget director Richard Darman.[3] They argued that the government should not interfere with the normal workings of the marketplace by picking industrial and technological 'winners and losers.' Hence, the Bush Administration's R&D budgets, which included relatively large amounts for basic science research and huge expenditures for military R&D, gave very little to commercial technology. The only R&D deemed worthy of government support was that tied to national missions like defense and health, or basic research, which the private sector will not fund at appreciable levels.

Until recently, the U.S. debate on S&T policy has not attracted much public notice. But there are signs that it could become a critical national concern in the years to come. In his 1992 State of the Union speech, President Bush alluded to the importance of S&T to the national welfare, highlighting his Administration's $73.6 billion R&D budget request for 1993. Aside from raising government spending for basic science research, Bush asked for funding for so-called "precompetitive" and "generic" technologies, and for initiatives to transfer technology from federally sponsored R&D programs (such as those at the national laboratories) to the commercial sector. The civilian side of R&D spending remained dwarfed by the $43.2 billion—59 percent of all federal R&D monies—allocated to military projects, including a whopping $5.4 billion for the controversial Strategic Defense Initiative.[4]

President Bill Clinton has been more outspoken about the need for industrial policies, and his S&T program promises to be more far-reaching. He is calling for revitalizing the nation's industrial base through investments in infrastructure, civilian R&D, education, and economic conversion (that is, helping military factories retool for the manufacture of commercial products).[5] Clinton has proposed taking funds from the defense budget to create a national information infrastructure, to build a national high-speed rail network, and to develop new energy-production and environmental-protection technologies. He also favors increased R&D spending for high-tech industry and the creation of both a new civilian R&D agency and a technical assistance network modeled on the Agricultural Extension Service. Finally, he wants to retrain and relocate displaced defense workers and offer loans to help small defense manufacturers convert to non-military production.

Whatever the merits of these specific proposals, Clinton at least deserves credit for addressing the three national challenges noted above. He correctly ties military cutbacks and economic conversion to strengthening the industrial base and to investing in the environment, infrastructure, and other national needs. Moreover, Clinton's ideas open the door to a wider national debate on S&T policy. They allow consideration of alternatives beyond the *laissez-faire* military industrial policy of the Reagan-Bush Administrations and even beyond the dual-use policies advanced by high-tech enthusiasts and their allies in Congress and in academia.

This essay examines the conceptual issues, criteria, and elements necessary to create an alternative S&T agenda driven by domestic national needs and implemented through expanded public participation. It begins, however, with a review and critique of the two main positions in the current debate—the military's industrial policy and the fashionable alternative of dual-use technology policy.

The Military's Industrial Policy

The military has dominated America's S&T policy since World War II. Although the U.S. defense budget and its R&D component are now being cut back, the military still pays for about one-third of the nation's overall R&D (which includes public and private R&D) and one-third of all R&D performed by private industry.[6] Massive expenditure on military S&T is a core element in the Pentagon's 'closet' industrial policy. For over forty years the Defense Department actively nurtured an enormous industrial base and a S&T establishment geared to the specialized needs of strategic and tactical warfare. Its policies spawned, among other things, a powerful aerospace, communications, and electronics (ACE) industrial complex.[7] The ACE industries received 80 percent of all federal R&D spending distributed to manufacturing firms in fiscal year 1989 (primarily from the DoD and the National Aeronautics and Space Administration, NASA), and were the largest performers of industrial R&D. Not coincidentally, ACE industries have been among the few in the manufacturing sector with a positive trade balance.[8]

The military-industrial sector also employs the nation's single largest bloc—from 20 to 30 percent—of scientists and engineers. Within the ACE industries, the most R&D intensive of all industries involved in durable goods manufacturing, scientists and engineers make up an unusually large share of the work force, ranging from nearly 20 percent to over 40 percent (in all manufacturing, in contrast, scientists and engineers make up only 5.5 percent of the work force).[9] Moreover, certain specific occupations—aerospace and astronautical engineers, electrical and electronic engineers, oceanographers, metallurgical and materials engineers, mathematicians and physicists—are especially concentrated in military-related jobs, largely in the ACE industries.[10] Other industrial sectors that have not enjoyed the government support accorded to the ACE complex, from traditional manufacturing (steel, auto, and textiles) to high-tech production (consumer electronics, civilian aircraft, semiconductor chips), have suffered major losses in market share to foreign competitors whose governments have comprehensive industrial and S&T policies.[11] For example, while European and Japanese firms and governments poured money into steel R&D in the 1970s, U.S. steel firms were cutting back and the U.S. government concentrated its R&D resources on military projects. In the early 1980s, when the White House was hawking the Strategic Defense Initiative program at $3-6 billion per year, the steel industry was refused $15 *million* for its Leapfrog Technology initiative, despite support from President Reagan's Science Advisor George Keyworth.[12]

Even the commercial product lines of companies that benefited from military spending are now losing ground internationally. Although U.S. companies lead the world in producing advanced military aircraft, missiles, and satellites, Airbus Industries, which is subsidized by four European governments, has captured wider shares of the commercial aircraft market and now threatens the predominant position of Boeing.[13] And while U.S. electronics firms excel in producing "Milspec"[14] electronic devices for military command, control, communications, guidance, and electronic warfare systems, Japanese and European firms, with help from their governments, have captured most of the consumer electronics and components markets, and are rapidly chipping away in the few remaining high-tech areas of U.S. commercial superiority—computers and semiconductors. Similar stories can be told for advanced materials, robotics, machine tools, supercomputers, and other dual-use technologies.

The end of the Cold War has forced DoD to bring its *de facto* industrial policy out of the closet, with cries from both inside and outside the Pentagon for more explicit military assistance for domestic industries. U.S. military planners are especially worried about the growing inability of U.S. firms to supply vital components and systems, which would increase reliance on foreign sources.[15] For example, the F/A-18 aircraft and Abrams tank are now dependent upon key Japanese and German parts. Defense analyst Jacques S. Gansler warns that for its next generation of precision-guided weapons, the DoD "may have to go to Japan or Europe to acquire the designs and technology it needs."[16]

But military and civilian technology requirements have significantly diverged over the years. Military technology, once a spur to commercial innovation, has fallen behind civilian technology in a number of crucial areas. For example, the civilian technologies of the fiber optics and microelectronics industries are far in advance of the military technologies.[17] Meanwhile, because of entrenched government practices, weapons manufacturers are having difficulty using the latest technologies available in the commercial sector.[18]

The reasons for these problems are institutional, administrative, legal, and technical. The civilian and military industrial sectors each have substantially different technology requirements, engineering approaches, and marketing practices. Success in commercial production hinges on rapid speed to market, low cost, and high volume. The design and production of advanced military systems, in contrast, involves very long product cycles, low-volume production, and the use of expensive, customized, highly specialized components and technologies. Military manufacturers must adhere to virtually thousands of product specifications and standards. Development lead times "have lengthened to the point of absurdity," Gansler observes, with the Pentagon on average taking more than 17 years to bring a new weapons system into production.[19] The Office of Technology Assessment reports that the armed services prefer to use older technologies, because "the acquisition system forces [managers] to assume responsibility when [a new] part or component fails, thereby jeopardizing their careers."[20]

Spin-off, Spin-on, and Dual-use

The problems afflicting military industries have prompted government leaders and industry analysts to propose substantial changes in the Pentagon's acquisition system. These reformers want to reduce the barriers between civilian and military industries by greatly expanding the use of commercially available, domestically made, state-of-the-art technologies and off-the-shelf components in weapons systems. This process is characterized as "spin-on"—in contrast to "spin-off," in which military technical innovations are adopted to civilian applications.[21]

The reformers want to overhaul the military's morass of unique specifications and regulations and to harmonize military industrial design and production with 'best practices' in the civilian high-tech sector—practices that emphasize low cost and high quality. They also call for increased DoD funding for manufacturing-related technology and for substantial federal investments in dual-use technologies critical to both defense and commercial competitiveness.[22] In response to a congressional request, defense planners came out in 1990 with a list of 22 such critical technologies.[23] This prompted several other government bodies (the Department of Commerce and the Office of Science and Technology Policy OSTP) and private-sector groups (the Council on Competitiveness, the Computer Systems Policy Project, and the Aerospace Industries Association) to prepare similar lists of dual-use technologies.[24]

The overlap between these lists is striking. Indeed, the doctrines of critical technologies, spin-on, and dual-use all reflect a convergence of interests between those who support a commercial high-tech industrial policy and those who seek Pentagon reform and a stronger military role in civilian manufacturing. High-tech policy proponents and members of Congress who have been unable to obtain desired levels of federal funding for civilian R&D agencies such as Commerce's National Institute for Standards and Technology (NIST)—let alone a new civilian technology agency established to carry the competitiveness ball—have been forced to turn to the one government entity seemingly immune from funding problems: the Pentagon.

As Gansler, a renowned military reformer, aptly explains, the advantage of such a civilian-military, high-tech marriage is that the DoD can "bring to commercial industry its very significant R&D funds, its ability to serve as the initial buyer, and its potential for capital equipment investments and labor training."[25] Often against the wishes of the White House, military reformers and high-tech industrialists (with some congressional support) have endorsed DoD initiatives in fields where civilian and military projects can both benefit, such as advanced materials, microchip manufacturing, computer-aided manufacturing, robotics, high performance computing, optoelectronics, high-definition display, advanced propulsion, and biotechnology.

Since 1980 the Office of the Secretary of Defense has supported the Manufacturing Technology Program ($138 million requested in fiscal year 1993), and since 1986 the Air Force has funded the National Center for

Manufacturing Sciences (funded at $37 million in 1992). But the principal DoD (and federal) sponsor of dual-use projects since the mid-1980s has been DARPA.[26] Since the mid-1980s DARPA has given the following grants: $100 million annually to Sematech, a consortium of semiconductor firms; $15-$20 million annually to the Software Engineering Institute at Carnegie Mellon University; over $150 million for high-definition television; over $1 billion for high-performance computing ($163 million has been requested for fiscal year 1993); substantial funds to other manufacturing technology programs ($165 million requested for fiscal year 1993); and $50 million (in 1991) for the recently initiated pre-competitive technologies program.

Through the Back Door

Despite substantial government financial commitments, the dual-use approach remains controversial. Many powerful individuals within the Pentagon and the White House opposed this strategy, as the 1990 firing of DARPA's director, dual-use advocate Craig Fields, attests.[27] But mounting public pressure to do something about the nation's flagging economic performance forced an accommodation with dual-use advocates. Indeed, by employing a piecemeal, 'back-door' strategy, high-tech policy champions secured congressional support and grudging acceptance by the Bush White House for their agenda.

One sign of White House softening was the 1990 OSTP report to Congress, entitled "U.S. Technology Policy." Leading dual-use advocate Lewis Branscomb from Harvard's Kennedy School quips that "building a consensus in the White House for any document with the words 'technology policy' in the title was no small achievement."[28] Backed by OSTP director D. Allan Bromley, the Bush Administration also endorsed "generic" pre-competitive R&D programs favored by high-tech enthusiasts, such as NIST's Advanced Technology Program and large-scale, interdisciplinary R&D initiatives successfully planned and pushed by the cross-agency Federal Coordinating Council on Science, Engineering, and Technology (FCCSET). These initiatives included the $665 million allocated in fiscal year 1992 and $803 million requested for fiscal year 1993 for the High-Performance Computing and Communications Program.[29]

Perhaps to mollify critics, the White House announced in 1992 a National Technology Initiative, spearheaded by the Department of Energy (DoE), that will facilitate technology transfer between the federal laboratories (including the weapons labs) and private industry.[30] Nevertheless, the Bush Administration's 1992 and 1993 budgets requested only paltry funds for commercial technology programs. At the same time, these budgets allocated most of the federal largesse to military R&D programs, especially large weapons systems. For example, more than half the proposed increase in R&D funding in the fiscal year 1993 budget request is for the Strategic Defense Initiative.[31]

Toward the end of the Bush Administration, influential private and public sector groups increased pressure for a national dual-use policy. For example, a report by the Carnegie Commission on Science, Technology, and Government recently urged the transformation of DARPA into a National Advanced Research Projects Agency (NARPA) to oversee the 'national technology base' under the Pentagon's auspices.[32] A National Academy of Sciences report, released in March 1992, recommended investing $5 billion for a quasi-government corporation to work with industry on developing "precommercial technologies," while affirming the role of DARPA in civilian technology.[33] Likewise, Congress reasserted DARPA's lead role in managing the nation's dual-use program in the FY 1993 defense authorization and appropriates bills.[34]

In contrast to its predecessors, the Clinton Administration has responded positively to the dual-use policy initiatives. President Clinton's economic plan would greatly expand spending on dual-use technologies. In addition, it affirms DARPA's central role in coordinating the nation's dual-use program. The "D"—for Defense—was also dropped from DARPA's name, reflecting its broadened mandate to support civilian technology initiatives. At the same time, the new ARPA will continue to serve as the Defense Department's primary R&D agency.

The Limits of "Dual-Use" Policy: A Conceptual Critique

One virtue of the current debate over national S&T policy is that it calls into question the two ideological pillars of the Cold War economy in the United States: the almost religious faith in market forces to drive economic progress, and the belief in military technology as a key catalyst for the development of civilian technology. Both pillars have dominated America's S&T policy for the past five decades. They rest on specific assumptions about the causes of economic change and technical innovation and about the appropriate role of the government in guiding S&T progress. But recent theoretical and empirical work from a number of economists and economic historians—most notably, Richard Nelson, Sidney Winter, Christopher Freeman, Giovanni Dosi, and Nathan Rosenberg[35,36]—casts doubt on these assumptions. A new school of thought, influenced by the writings of Joseph Schumpeter, has emerged challenging the *laissez-faire* premise that natural market forces will yield socially optimal allocations of R&D resources.

The Limits of *Laissez-Faire*

According to neoclassical economic theory, the managers of firms invest in R&D and new technologies to maximize profits. Resources are allocated efficiently when supply meets demand in a competitive market at an equilibrium price. In this ideal world the best economic and social results are attained if R&D allocations are left to the decisions of private firms. This explains the free-marketeer injunction against the government interfering with this natural process by "picking winners and losers" among technical and industrial alternatives.

Dual-use proponents challenge the *laissez-faire* model. While they do not contest the importance of markets as the primary mechanisms to allocate resources, they contend that non-market, institutional forces—especially deliberate government policies—also are needed to guide technical innovation and to improve the nation's industrial performance. Writing earlier this century, Joseph Schumpeter argued that technical innovation is the principal engine of economic progress in capitalist economies. He demonstrated that orthodox equilibrium market theories cannot explain how technology is developed nor how it shapes economic competition in capitalist societies. On an abstract level, Schumpeter's view—and that of his latter day, dual-use followers—rebuts the assumptions of traditional theory. The assumption of profit maximization is predicated on firm managers having access to perfect information about markets and available technology, and on their ability to choose the best of these options to maximize efficiency and quality in production. In the real world, however, information is necessarily imperfect and uncertain. Organizational, institutional, and technical constraints limit the ability of managers to maximize the returns on technological bets. The dynamics of technological innovation and diffusion in this imperfect world also create disequilibria; features inherent in the process of technical innovation and the organization of private firms precipitate "market failures."

Adjustment mechanisms in technological competition differ radically from the linear allocation mechanisms posited by neoclassical models.[37] Neo-Schumpeterian economists put forth an "evolutionary" model for explaining complex interactions between technical change and market dynamics. Recognizing the inherent uncertainties and risks involved in technological innovation, private firms place their technological bets by engaging in a process of "search" (using R&D and market research) and "selection" (among new technologies).[38] But market signals alone are often "not sufficiently specific," as Nathan Rosenberg notes, for private firms to produce economically and socially optimal R&D investment portfolios.[39] Whether a firm's technology investments pay off depends not only on market forces but also on a variety of institutional and social factors. These factors together operate as "selective devices," or in Rosenberg's terms, as "inducement" and "focusing" mechanisms, which move technological change along specific development paths.[40]

Less abstractly, the proprietary nature of technological knowledge is one of the major constraints on the capacity of firms to make optimal R&D choices. To protect and guarantee a sufficient return on R&D investments, firms try to control access to critical technical information through patents, copyrights, internal security measures, and other devices. Economic success depends on the ability of firms to appropriate the results of such investments. But when proprietary rights are not desirable, or not possible, government intervention may be necessary.

Certain types of technological information and innovation however are not easily appropriated by private companies for competitive advantage. It is generally conceded, for example, that the advancement of basic scientific and engineering research requires an open exchange between researchers in the field. The results of such inquiry usually are freely available in unrestricted professional publications. Moreover, certain kinds of technological innovations and products that can be easily copied by competitors or users are inherently difficult to protect with patents or copyrights. This is a sensitive issue within the computer software industry.[41]

These limits on private appropriability of technical knowledge inhibit private investment in R&D. Why should a firm invest in a new technology if a competitor or a downstream user may be able to derive an equal or greater benefit?[42] The problem is compounded by the escalating costs of conducting research in certain areas of advanced science (high-energy physics, superconductors) and advanced technology (semiconductor manufacturing). Without government support, only the largest and most affluent private firms like IBM and AT&T can afford to foot the bill.

Whither Industrial Policy?

According to neo-Schumpeterian economists, the obstacles to private sector R&D investment cannot be removed through ordinary marketplace incentives. They represent serious sticking points in the traditional *laissez-faire* model of economic change. For high-tech policy proponents, these market failures justify government intervention in allocating S&T resources. They contend that direct subsidies, tax incentives, and other inducements for private sector investment are needed to bolster economically strategic S&T areas. The government could help level the playing field for competing firms by spreading the costs of R&D among them. It could support cooperative industrial R&D projects by allowing firms to enter consortia (without fear of antitrust violations) or joint ventures with academic institutions and government laboratories. Collaborations of this nature reduce unnecessary duplicative efforts and enable more companies to have access to a common technical knowledge base.

Even the Reagan and Bush Administrations have accepted the need for government financing of high-risk, long-term basic research by the private sector, especially if it is relevant to the missions of specific federal agencies like defense and health. The free-market objection has been directed against government funding for commercially relevant R&D. High-tech policy proponents have been quick to respond that the targets for federal R&D should be set by industry, not by government bureaucrats. In addition, they underscore that the R&D agenda should stress generic, "pre-competitive" technologies that benefit several, rather than individual, industries and firms.[43] Although these caveats have been made largely to mollify the free-marketeers, high-tech industrialists themselves have concerns about government practices that may give undue advantage to competitors. They want a partnership with government that would guide S&T along commercially advantageous paths, but also one that would further protect their proprietary rights regarding technical innovation.

Regardless of the terms in which the high-tech proposals are couched, they constitute a conscious shift in U.S. economic policy away from *laissez-faire* to industrial policy.[44] Free-market ideologues have opposed this industrial policy since the Eisenhower Administration, when they derailed the bid of the Commerce Department's National Bureau of Standards (now NIST) to become the lead civilian technology agency in the United States.[45] In the 1984 presidential campaign President Reagan similarly beat back the labor-oriented industrial policy of the Democratic Party, which was pitched as a way of reviving heartland industries (automobiles, steel, textiles) that were decimated during the late 1970s and early 1980s.

In today's economic and political climate, however, new high-tech, dual-use industrial policy initiatives may not be so easily dismissed. The question is no longer whether the United States should have an industrial policy, but rather, what kind of industrial policy? As Richard Nelson observes, the emphasis on private-sector motives and behavior in much of the discussion "leaves hidden in the shade the considerable long-standing public involvement in high-technology industries" in almost every advanced industrial nation, including the United States.[46] As is well known, Japan and many Western European nations have implemented ambitious industrial policies over several decades, carefully supporting the development technology for desired commercial and societal objectives like energy efficiency, advanced modes of transportation, and enhanced manufacturing performance.[47] Large-scale national initiatives being launched by our competitors include the European Airbus (commercial aircraft), RACE (telecommunications), ESPRIT (information technology), and BRITE (manufacturing) programs, as well as Japan's Fifth Generation Computer Program.[48]

Neoclassical hand-wringing over U.S. industrial policy is especially hypocritical, given the federal government's major role in guiding industrial development for national goals, often with the blessings of free marketeers, throughout the country's history. Beginning in the mid-1800s, agriculture was one of the first economic sectors to benefit from public programs such as land-grant colleges, R&D support, experimental stations, and technical extension services. The National Advisory Committee for Aeronautics (NACA, which was later absorbed into NASA) was created in 1915 in response to the nation's perceived inadequacy in aviation research and production. By World War I the United States had only 23 aircraft, while the French had 1,400 and the Germans 1,000. NACA's purpose was to advance aeronautical S&T and to advise the military and other federal agencies on aeronautical R&D.[49] NACA (and later NASA) not only supported research important for the industry but also underwrote much of the research infrastructure associated with innovation in airframes and engines.[50]

According to Linda Cohen and Roger Noll, since World War II every Administration has pursued policies directly benefiting select areas of industrial and technological development.[51] The Roosevelt and Truman Administrations funded the beginnings of the computer industry. The Eisenhower Administration supported the semiconductor and nuclear-power industries. The Kennedy and Johnson Administrations gave the nation the Apollo program, the supersonic transport (SST), and the war on cancer. And Presidents Nixon, Ford, and Carter spent billions on energy R&D in response to oil crises of the 1970s. Even President Reagan supported commercially attractive programs for developing breeder reactors, nuclear fission, orbital manufacturing facilities (Space Lab), and rocket planes (the Orient Express).

For many years the United States also has conducted a medical industrial policy, spawning a medical-industrial-academic-government complex every bit as entrenched (if not more so) as the military-industrial complex. It has garnered a set of powerful vested interests with a stake in maintaining the current system including: medical professionals, hospitals and clinics, pharmaceutical and medical-product firms, academic research centers, government agencies, and insurance carriers.[52]

In almost all these cases government intervention helped spawn a select set of industries and develop the S&T base needed to pursue national objectives. Government policies also helped to shape specialized market and industrial structures, specific types of technology, and new patterns of public and private investment. However, no government policy since World War II has had as significant an impact on the country's economic, industrial, and S&T development as its military-industrial policy.

Beyond Spin-Off

The full extent of the Pentagon's industrial policy and its economic consequences have been elaborated elsewhere.[53] Of the roughly $300 billion of annual defense spending, more than 40 percent is for procurement of products or for R&D,[54] paid mostly to firms in the manufacturing sector. Of the over 200 industries responsible for 95 percent of defense production, over 60 depend on the military for 25 percent or more of their sales. Another 40 industries, mostly suppliers of components like fasteners, ball and roller bearings, and industrial controls, get between 10 to 25 percent of their business from military contracts. The shipbuilding, ordnance, and ammunition industries are the most defense dependent, with over 70 percent of their sales derived from military spending.

But the largest portion of military procurement and R&D dollars—nearly 80 percent—goes to the ACE industries discussed earlier.[55] Throughout most of the Cold War, the U.S. government granted the ACE industries advantages enjoyed by few others in a free-market economy. Among other things, it directly provided capital, equipment, and facilities, it bailed out troubled firms through loans and subsidies, it limited competition, it guaranteed profits through preferential procurement practices, and it provided special export supports to bolster foreign arms sales. Perhaps most importantly, the military subsidized the ACE firms' R&D activities at massive levels, enabling them to develop new products with little risk to themselves. As noted earlier, the ACE industries receive the overwhelming bulk of federal R&D funds earmarked for manufacturing firms. The military agencies also underwrote the costs of an enormous S&T base for weapons development, including nearly 200 government laboratories and research programs at hundreds of academic, industrial, and non-profit institutions.

To justify high levels of military spending, defense proponents tout the numerous economically important spin-offs from these tremendous investments—commercial jetliners, computers, microelectronic devices, and advanced composites, to name a few. Critics argue just the opposite: it would have been more economically efficient and socially beneficial had we invested directly in civilian technologies. But the spin-off argument is being laid to rest, not by long-standing opponents of arms spending, but rather by the proponents of dual-use industrial policy. For example, spin-on policy advocates Michael Borrus and John Zysman assert: "[M]ilitary spending and military technology development are not going to rescue the civilian economy from its competitiveness problems. Nor can they assure sufficient national technological development even for security purposes."[56]

The arguments for moving from a spin-off to a spin-on orientation in military-procurement practices, bolstered by federal support for R&D in dual-use technologies, include the following: the requirements for civilian and military technology are diverging; military S&T is lagging behind civilian S&T; and the dependence on foreign sources for high-tech components is growing. U.S. military-industrial policy has allocated most of the nation's S&T resources to development paths that have yielded decreasing returns to the civilian sector. Meanwhile our principal trading partners, Japan and Europe, have nurtured their S&T base to strengthen domestic industrial performance.

The neo-Schumpeterian economists note that technological change tends to be constrained along very specific paths. These paths are defined by internal technical possibilities and limits, by market characteristics, and by other economic, institutional, and social conditions. Technology development and, to a lesser extent, science research are "path dependent" processes. Their trajectories are contingent upon the workings of non-technical forces. Hence, in Giovanni Dosi's terms, by providing financial support to R&D and guaranteeing public procurement over the past decades, the U.S. "military and space programs operated . . . as a powerful *focusing mechanism* towards defined technological targets."[57] With its intricate network of prime contractors, subcontractors, and suppliers, its specialized markets, and the focused procurement and R&D policies from its government patron, the military-industrial system comprises a unique "technological paradigm." It moves technological innovation and diffusion along paths that are most relevant to its requirements but not necessarily optimal for meeting the needs of the civilian sector.

Dual-use and spin-on technology policies are meant both to reduce the influence of the military on the trajectories of civilian technological development and to foster a new civilian-military industrial integration, in which civilian and military high-technologies are developed in a complementary and mutually-reinforcing process. Military dollars can support the development of civilian technology without skewing or slowing commercialization of technical innovation, while the military can get domestically available, high-quality products at lower costs. Not coincidentally, new commercially friendly procurement practices by the Pentagon also would expand domestic markets for commercial products.

The Limits of Dual-Use

Dual-use advocates have effectively challenged both the *laissez-faire* and military-industrial policies that have guided the development of America's S&T. Moreover, they support investments that undoubtedly will yield some economically valuable technical innovations. And they are pushing for a shift to spin-on procurement practices that can reduce military spending while helping domestic high-tech suppliers.

Nevertheless, dual-use policy is not socially optimal for two reasons.[58] First, despite intentions to the contrary, dual-use policy will perpetuate, if not strengthen, the military's influence on the nation's S&T development. Second, it precludes investments in vital social, economic, and environmental needs.

Maintaining Pentagon sponsorship of high-tech development will hinder technology transfer in significant areas of commercial and national needs. Despite substantial, long-term DoD funding, U.S. firms marketing dual-use technologies, such as robotics, semiconductors, supercomputers, superconductors, airframes, advanced propulsion, and advanced ceramics, continue to lose ground to Japanese competitors. One reason is obvious—the widening gap between military and civilian paradigms in technological development. In addition, not all critical areas of civilian S&T are relevant to the military. In the final analysis, DARPA, the dual-use technology leader, is a military agency with a military mandate. Its main clients are the military services and intelligence communities. Most of its projects are military-related, and its main contractors are well-known weapons manufacturers. Hence, of necessity, its programs favor the needs of military over civilian producers.[59]

In any event, dual-use initiatives make up only three percent of the defense R&D budget. Over 90 percent of the Pentagon's $43 billion R&D commitment is devoted to military projects linked to procurement, with little relevance to commercial production. This priority will continue to over-shadow investments in civilian research. To make a real difference—that is, to bring U.S. civilian R&D spending as a share of GNP up to the level of Europe and Japan—would require a shift of $13 to $16 billion from defense to civilian sectors, assuming a matching investment by the private sector.[60]

Dual-use policy is based on the "trickle-down" notion that if the competitiveness of our "strategic" high-tech industries is strengthened, the spill-over will benefit the rest of the economy. But there are no specific mechanisms for getting bottom-up guidance on directing these spill-overs to the areas of greatest economic and social need. Despite a large body of evidence that worker participation in technology decisionmaking improves productivity, labor plays little role in setting the dual-use strategy.

In fact, dual-use policy retains, if not tightens, the control over the U.S. agenda for R&D by an elite group of industrial, governmental, and professional leaders who have little incentive to incorporate broader social or environmental concerns.

Dual-use policy does not provide adequate demand-side inducements to promote new directions for S&T. It assumes that global market forces will provide the necessary incentives for commercial innovation, if only the federal government will help out on the supply side. Given the global recession and international instability, this seems like a dubious premise. A policy based on "competitiveness" goals alone will not provide the demand-side inducements needed for promoting new directions for technical change. Economists David Mowery and Nathan Rosenberg observe:

> [B]enefits that are sometimes perceived to flow from military R&D are in fact the product of military R&D plus frequently massive military procurement.... Without the pull of defense procurement in such sectors as jet aircraft, integrated circuitry, and computers—especially in the critical early years of development of these technologies—the impact of military R&D spending alone would have been far smaller.[61]

Similarly, without comparable public investments in civilian needs like alternative energy and mass transit to create new "demand-pulls" to guide S&T development, the military will remain the demand creator of last resort for advanced technologies.

In sum, a dual-use agenda that invests only in a handful of so-called "critical" technologies, based on narrow economic and military criteria, will limit opportunities for economically and socially vital technologies. Although rebuilding public infrastructure, developing renewable energy sources, and cleaning up the environment would enhance the nation's economic performance and stimulate new species of technical innovation, dual-use policy does not support such programs. In fact, with a tightening federal budget, dual-use programs may well preclude investments in these critical areas of national need.

Toward a Third Way

The preceding critique raises pragmatic concerns regarding laissez-faire and dual-use policies, but the principal issues are normative. Neither strategy adequately addresses critical economic, social, and environmental problems. Moreover, both are elite-controlled and trickle-down approaches.

They do not guarantee an equitable distribution of S&T resources and benefits to many important constituencies, particularly workers, minorities, women, and the poor. Worse, it may become harder for such groups to obtain the financial and technical support they need to prevent or redress harmful "externalities" of S&T, such as public and occupational health hazards caused by toxic wastes in their homes, communities, and workplaces.

Is a third way possible? Can an alternative industrial policy be constructed that is driven by national needs? Can the national mandate in S&T policymaking be broadened to include social, economic, and environmental interests?

One preliminary challenge that hinders the design of another S&T policy is that American culture greatly mystifies, if not mythologizes, S&T. The forces of S&T are portrayed either as supremely beneficent and capable of saving us, or as demonical and capable of threatening our survival. On the one hand is Prometheus, the mythical giver of fire to humankind, the symbol of S&T as the source of human progress. On the other hand is Frankenstein, Mary Shelly's fictional character who has become a modern archetype of technology run amok.

A new, normative-based S&T agenda must move beyond this disempowering polarization. It is useful to enunciate a few basic principles. First, the forces of S&T are neither uncontrollable nor autonomous. Human institutions and choices guide their development and application. Second, S&T is not value free. The goals, objectives, and criteria used in guiding the development of S&T matter. Third, the political, social, and economic interests participating in S&T decision-making also matter. Finally, it is better to influence the S&T development cycle at its front-end than to regulate its adverse consequences at the back-end.

Human Choice and the Control of Technology

The forces of S&T can and must be guided by conscious human agency. But in popular culture they are largely perceived as outside of human control, or at best, the domain of a small technocratic "priesthood." For example, Lisa Gallatin, director of the Boston-based Coalition on New Office Technology, observes that for most of the rank-and-file clerical and service workers who are members of her organization "technology is not a bread and butter issue."[62] Despite the obvious impacts of technology on their jobs, she says, "the idea that technology can be controlled is very far removed from their day-to-day lives." Even S&T practitioners accept the notion that science and technology are independent forces. Scientific researchers frequently espouse the autonomous, neutral nature of their work. For example, Michael Dertouzos, director of the Massachusetts Institute of

Technology Laboratory for Computer Science, referring to the effects of government patronage on research, argues that "science, especially the science yet to be discovered, isn't too happy going where we want it to go. It goes where it wants to go."[63]

Recent research in the social studies of science seriously challenges this view. The work of scientists—the selection of research problems, the methods of inquiry, the modes of analysis and interpretation, and even the research results—is a product of social, political, and economic factors, such as the sources and purposes of funding.[64] Technology, which involves the application of scientific knowledge, has even fewer claims to autonomy. Technology may be defined as how we produce the things we need.[65] It encompasses hardware and software, the codified knowledge and know-how of technical workers, and the organizational and institutional forms that shape how technical resources are employed. As the neo-Schumpeterians contend, technological innovation must be treated as an endogenous process, driven by institutional and market factors.

Goals Matter

Science and technology are not value free either. The goals and objectives of those who pay for, produce, and apply S&T carry weight. Both corporate managers and government agencies attempt to impose their own priorities and criteria on the R&D process. Corporate managers typically worry about cutting costs, maintaining product quality, and controlling employee outputs. They rarely consider occupational or environmental criteria in S&T decisionmaking, unless regulation or collective bargaining forces them to do so.

America's R&D agenda has been dominated by military objectives, at the same time as Japanese and European governments and corporations have targeted most of their S&T investments at civilian manufacturing. U.S. scientists and engineers have concentrated their creative power in producing ever more exotic and expensive military hardware, while their Japanese and European counterparts have been guided by the goal of getting high-quality, low-cost goods to market as quickly as possible. Other governments have set national goals, such as energy independence, to focus the development of new technology. Susan Walsh Sanderson describes how the Japanese consumer-electronics companies, with government assistance to create "precompetitive ventures to work on basic problems in integrated circuits to television production," shifted to solid-state electronics, in part "because of their intense concern for reducing power consumption after the first oil crisis." Meanwhile, the U.S. government "had no coherent energy policy at the time and certainly no coherent policy towards the electronics industry."[66]

Another example of demand-pull policy is the Japanese Magnetically Levitated (Maglev) Train Program, a part of Japan's commitment to high-speed mass transportation.[67] Maglev supports not only advanced transportation technology but also basic research. According to an Office of Technology Assessment study on superconductivity, "the project . . . illustrates the way in which applications drive technical development at more fundamental levels."[68] To stay in the forefront of maritime technology and to upgrade their merchant marine fleets, Norway, Japan, and West Germany have sponsored "Ship of the Future" projects. These efforts, Tom Forester observes, have created a demand for corrosion-resistant and maintenance-free deck equipment made of ceramics, plastics, and metal alloys that can withstand the salty rigors of life at sea."[69]

Participation Matters

Those who participate in S&T decisionmaking will have the greatest ability to impose their goals, values, and criteria on S&T development. Currently most S&T decisions are made by a small, elite group of scientists, engineers, corporate managers, and government officials. Others have little or no say in these arcane deliberations, even though these decisions greatly affect their lives.

To forge a S&T agenda reflecting broad social and environmental goals, the political-industrial-scientific elite must be challenged to open up the nation's S&T policymaking process to many more sectors of society. Democratic participation must be a primary goal. If S&T policies are to address the broad concerns of society, then the constituencies associated with these concerns must be represented in the policymaking process. Is it no coincidence that shortly after Bernadine Healy came on board as director NIH established a new Women's Health Initiative?[70]

Starting at the Front-End

A democratic S&T agenda requires injecting social criteria in the R&D and design stages of production, where most of the crucial decisions about materials, means, and methods are made. Historically, efforts to redress S&T abuses have focused mainly on regulating or stopping technologies after the damage has occurred. Society has sought to enforce standards on the design and production of technologies by punishing abusers through fines and legal actions. A less costly approach is to emphasize "prevention," by setting desired criteria and objectives at the front-end of the S&T development cycle.

In the fields of environmental protection and occupational health and safety, for example, preventive practices are beginning to be applied.

Promoters of sustainable development are now designing production processes, technologies, organizations, and final products to minimize pollution and the use of resources.[71] Similarly, designing machine tools both to enhance the skills of workers and to expand the participation of labor in the production process would increase productivity and reduce the job-displacing impacts of new technologies.

A New S&T Agenda: Targeting National Needs

What would a democratic strategy for linking S&T investments to vital national social, economic, and environmental needs actually look like? Such a strategy would contain five principal elements: new priorities to mobilize national resources; public investments to support these priorities; economic adjustment programs for businesses, communities, and workers, especially to assist in the transition from military to civilian production; new S&T initiatives; and new modes of democratic participation in S&T decisionmaking.

New National Priorities

The United States needs new national priorities to guide public and private sector commitments in the coming decades. Just as national security served as the primary goal for U.S. industrial policy during the Cold War, now three other explicit goals must guide government action — economic stability, environmental cleanup, and decent health care.[72]

Economic-stability policies would strengthen the well-being of workers, businesses, and communities. They would improve the nation's industrial performance, stimulate new economic growth, and create decent jobs, while at the same time preserving the integrity of communities and increasing job security for workers. In their pursuit of profits, corporations (and government planners) all too often ignore the destructive effects of plant closures or employment cuts on households, local governments, and small businesses.[73] With one of the least comprehensive economic-adjustment programs in the advanced industrial world, the United States will be unable to guarantee economic stability until it can protect those most affected by economic loss.

Environmental sustainability is a critical condition to achieve economic security. The attention garnered by the 1992 Earth Summit in Rio de Janeiro, Brazil, underscored that environmental protection has become a global concern. Governments everywhere are being forced to invest in cleanup and to strengthen environmental regulations. However, as most ecological problems derive from uncontrolled production practices, governments also must integrate environmental objectives and criteria into their plans for economic development.

Americans will neither be nor feel secure until their health and safety is better protected. But as medical costs escalate, more and more U.S. citizens are losing access to adequate health care. A new national health policy is needed emphasizing prevention, equal access to affordable and high-quality medical care, and reduced threats to health and safety in the workplace and in communities. These objectives must be incorporated not only into the nation's public health-care delivery programs but also into health-and-safety regulations and health-related R&D investments.

Finally, the emphasis of industrial policies must be shifted from national security to global security, so that the security and well-being of all people in the world is promoted. The United States and the other industrial powers have the responsibility and the resources to help less advantaged societies achieve real security. Global cooperation, instead of national competitiveness, must become the watchword for action. Competitiveness raises the specter of xenophobic competition, which, as MIT economist Paul Krugman observes, "like national defense, . . . can be easily used as a patriotic cloak for special interest politics."[74]

New Public Investments

Achieving all these goals will require new public investments in several critical areas that have suffered from government neglect over the past decade. The list of worthy national expenditures is a long one and includes rebuilding public infrastructure, cleaning the environment, developing renewable energy sources, upgrading schools, revamping health care, constructing mass transit, and building affordable housing. Many economists believe that public investments would stimulate the productive base of the U.S. economy by increasing private investment, productivity, and profits. Jeff Faux and Todd Schafer of the Economic Policy Institute report that "the United States could not have successfully developed a powerful private economy without large and sustained government outlays in transportation, education, and the generation of new technologies."[75] The United States invested less in these areas in recent years, while other industrial nations invested more, "setting the stage for further declines in living standards and competitiveness." Faux and Schafer calculate that the United States currently suffers an "investment deficit" of between $60 and $125 billion each year.

Economic Adjustment Programs

Increased public investment requires redirecting financial and technical resources from the military to the civilian sector.[76] This is one of the fundamental challenges and the opportunities the United States faces as it dismantles its Cold War economy. The government needs to implement economic-conversion and adjustment-assistance programs to help military-dependent communities and workers make this transition with the least pain. Capital and technical resources from the national government are necessary to retrain workers, to convert military factories, and to diversify local economies. Federal and state economic-assistance programs that serve workers also must be strengthened. Such programs represent an investment in the efficient re-use of economic resources, as well as a means to reduce the costs to society of human hardship.

Science and Technology Initiatives

New civilian S&T initiatives coupled with demand-creating public investments would shift patterns of U.S. technical innovation in directions compatible with socially needed goods and services.[77] Each initiative would target a unique set of cross-cutting, interdisciplinary R&D programs geared to a specific social need. Generally, these initiatives would have four characteristics.

First, each initiative would be guided by broader social criteria than those espoused by dual-use policies. Federal R&D projects no longer would stress only arcane military specifications or commercial objectives like cost, productivity, product performance, and marketability. They would also stress other goals like environmental sustainability, energy efficiency, occupational health and safety, job security, worker autonomy and participation, equity, and access.

Second, these initiatives would build the S&T base needed to support new public investment programs, just as most S&T over the past forty-five years was built up to support national defense. The programmatic elements would include many traditional S&T policy instruments, such as direct subsidies, R&D tax credits and incentives, and public-private R&D collaborations. The types of activity supported would include long-term basic and applied research, generic technologies, systems prototypes, technology test-beds, and the design and testing of products tied to direct government procurement.

Third, these initiatives would target a wide range of S&T fields, including most critical technologies favored by dual-use advocates, such as advanced materials, microelectronics, and advanced computing. However, the new socially-based criteria also would focus so-called "generic" technologies along different paths of innovation. Semiconductor manufacturing technologies that minimize toxic chemical use, for example, differ substantially from comparable technologies developed only according to narrow commercial or military specifications. At the same time the initiatives would select areas of research not on the many critical technologies lists that have been generated by pubic and private sector groups. So far, only an OSTP report identifies "*national* critical technologies" that address broader societal needs, such as environmental protection, energy efficiency, and transportation.[78]

Finally, these initiatives would expand the portfolio of R&D opportunities for the research community. They would most likely be coordinated across several agencies, similar to the high-performance computing initiative orchestrated by OSTP and FCCSET. These programs would give new direction to in-house and extramural programs of federal R&D agencies, but not necessarily hurt their budgets. For example, were the DoE to change its mandate from nuclear energy and nuclear weapons to sustainable energy, the national laboratories under its direction—especially, Livermore, Sandia, and Los Alamos—could more easily retool to serve the more socially beneficial agenda. Other agencies such as the Departments of Transportation, Commerce, and Labor, or the Environmental Protection Agency (EPA) might even find their R&D budgets beefed up. Proposals for new agencies, such as a National Institutes for the Environment or a civilian advanced technology agency, would have a better chance of success. DARPA's role in civilian technology, however, would diminish, since it would no longer be the primary sponsor of cutting-edge research—even dual-use research.

The Grand Challenges

The United States needs to meet the new "grand challenges" in S&T, similar in scope to past efforts to reach the moon, to cure cancer, or to produce a strategic defense system. The leading challenges include sustainable development, workplace quality, transportation infrastructure, revitalizing manufacturing, and rebuilding cities.

Sustainable Development

Can the United States produce industrial and agricultural technologies that are energy efficient and environmentally clean? Ken Geiser at the University of Massachusetts-Lowell notes that the notion of sustainability first appeared in the field of agriculture and is now being applied to manufacturing.[79] Sustainable agriculture aims to reduce the use of chemical pesticides and fertilizers, replacing them with practices that work in harmony with natural ecological cycles. Sustainable manufacturing emphasizes appropriate use of technology, environmentally benign material inputs, minimization of waste, energy efficiency, a safe working environment, and resource conservation.

Also key for sustainable manufacturing are technologies for pollution control and waste reduction that are designed into the front-end of production. While the U.S. spent $100 billion (in 1986 dollars) on pollution clean-up in 1990, less than one percent of this amount went to pollution reduction.[80] Increased spending on sustainable production technologies could benefit both the environment and the economy.

The Japanese and Europeans already are aggressively developing industries to serve the growing market for technologies, products, and services to protect and clean up the environment. By the end of the decade, this may become a $300 billion business.[81] There is also a growing market for alternative energy technologies. A sustainable energy-technology initiative might include renewed federal support for fuel cells, photovoltaics, a solar car, and conservation technologies, all of which were drastically cut in the Reagan years.

Quality of Worklife

Can the United States design technologies that enhance job security and the quality of worklife, including the health, safety, skills, participation, and autonomy of workers, and at the same time improve productivity and product quality? Developing technology for a better worklife would integrate "participatory design" and "skill-based automation" with efforts to reduce special hazards and stress-related work injuries. Pioneering work in this area has already been done in Europe. Professional and labor groups in the United States also are expressing growing interest in skill-enhancing technologies, such as those involving computers in manufacturing and service industries.[82]

Transportation Infrastructure

Can the United States build safe, efficient, environmentally benign large-scale transportation systems for both travel within and between major metropolitan areas? Much more investment is urgently needed to build new surface mass transportation systems and to improve the country's air and maritime transport capabilities. Systems are needed that not only can handle more people and cargo but also can go faster, more safely, and with less fuel, pollution, and noise. There are many new opportunities for technological innovation in transportation, including electric vehicles, high-speed rail, guidance control and tracking systems, and intelligent vehicle and highway systems.[83]

Revitalizing Manufacturing

Can the United States design new manufacturing technologies that satisfy the criteria of sustainability and quality of worklife, while strengthening manufacturing performance both in the high-tech industries and in traditional sectors like steel and textiles? Investments here might include support for advanced specialty steel manufacturing,[84] or "intelligent" textile machinery, as well as for leading-edge technologies like computers and microelectronics. Other investments might go to high-tech innovations applied to traditional industry and high-tech manufacturing processes that simultaneously enhance productivity and reduce toxic pollutants.[85]

Rebuilding Cities

What technologies does the United States need to rebuild transportation, communication, sewage, housing, and other public-service systems in metropolitan areas? Cities must be redesigned from a holistic planning and environmental perspective, one that looks at the interaction of infrastructure systems with economic and social functions. Initiatives here would especially target interdisciplinary R&D projects involving urban planners, economists, social scientists, and engineers.

Modes of Democratic Participation

James C. Peterson writes that "citizen participation is nearly synonymous with American democracy. . . . Few values in American society are held more deeply than the right of citizens to participate in decisions on matters that directly affect them."[86] Opening up S&T policymaking to include underrepresented constituencies such as workers, environmentalists, women, minorities, and small business people is essential for crafting effective and equitable industrial policies. Broader representation would promote different criteria for technical design and R&D-resource allocations.

Expanding the S&T Advisory Process

Public interests should be represented on the numerous advisory panels that help federal and state governments set R&D priorities. This would ensure that social and environmental criteria get incorporated into R&D budget, program design, and award-selection processes. Citizen advisors could serve on S&T policymaking bodies in both the executive branch (OSTP, R&D agencies) and in Congress (the Office of Technology Assessment, House and Senate committees). Similar advisory functions could be established at the state level and for publicly funded R&D ventures. Public advisors would not preempt the legitimate role of scientific experts to advise on matters of scientific merit and promise. In many instances, as in the environmental and occupational health areas, however, a single individual can represent both technical and public interests.

Access to Information and Expertise

Public access to technical information and expertise needs to be greatly expanded. Expertise, Peterson notes, "is a potent resource, and the public can ill afford to let utilities, developers, and corporations purchase a virtual monopoly of technical expertise."[87] The movement for science in the public interest of the 1970s made public access to scientific advisers in government and industry a central theme.[88] Responding to this movement, the Freedom of Information Act opened the files of governmental agencies, and the National Environmental Policy Act forced the airing of environmental concerns in publicly funded projects.[89] Today workers and citizens need much stronger "right-to-know" laws aimed at the private sector and public-private enterprises. It would also be useful to revive programs like the National Science Foundation Science for Citizens Program, which supported community-based S&T service centers and increased citizen involvement in science policy.[90] Similarly, the United States could establish a national technology extension service, modelled on the successful agricultural extension system. Local industries often need technical assistance, not advanced R&D, for upgrading operations, as the North Carolina-based National Apparel Technology Center recently discovered.[91]

Worker Participation

Both labor and business have an interest in greater worker participation in technology planning. The technological revolutions in information processing, communications, materials, and biotechnology are transforming the nature of work in modern societies. These new technologies represent both threats and opportunities for workers' job security, skill levels, autonomy, health, and safety. Over the years the International Association of Machinists (IAM), the United Auto Workers (UAW), and the Communication Workers of America (CWA) have pushed for a greater say in technology decisionmaking in collective-bargaining agreements. One especially important initiative to shore up the rights of workers to control technology occurred in the 1980s, when the IAM introduced (with little success) legislation based on a ten-point New Technology Bill of Rights.[92] The AFL-CIO's Industrial Union Department recently held conferences on technology and work that emphasized a stronger role for labor in shaping U.S. technology policy.

There is a wide consensus among business leaders that a highly motivated, high-skilled work force, coupled with greater participation in industrial decisionmaking, will yield substantial gains in productivity and higher quality products. Knowledge gained either tacitly or experientially in the production process is critical to technological innovation, especially in modern high value-added industries. As an Office of Technology Assessment study notes, progress in industrial production "depends heavily on experience and empirical know-how." The same study also concludes that "reluctance among American engineers and managers to learn from shop-floor employees hurts productivity and competitiveness."[93]

Worker participation schemes have drawn criticism from both for managers and labor unions. Managers, particularly at the middle-management level, have felt their traditional prerogatives threatened by worker participation plans, while labor unions often have believed, not without justification, that companies will use such efforts to undermine the power of trade unions. But the much touted Saturn auto plant satisfied both managerial and labor concerns and included the work force in every aspect of its development.[94] This and other success stories suggest that workers should genuinely be able to influence key decisions concerning technology and work organization, and they should be brought into the planning process at the outset.

Education, Training, and Employment Opportunities

Strong public education is a critical condition for greater citizen participation in S&T policymaking. It is not just a matter of improving the public's scientific literacy or S&T skill levels. Critical historical and social perspectives must be built into curricula, especially into technical courses. College engineering courses, for example, might stress problems relating to the environment or energy alternatives (e.g., solar cars) instead of the aerodynamics of fighter aircraft or the mechanics of landing a rocket on the Moon. This process has begun in some academic institutions, but not to any great extent.[95]

More important are greatly expanded S&T educational and employment opportunities for women and minorities. Women, African-Americans, and Latinos are grossly underrepresented in the science and engineering work force.[96] National S&T policies should prepare women and minorities for technical jobs but also should give them a greater entrée into the S&T policymaking system.

Getting from Here to There

Selling this program will not be easy. Nevertheless, there are some encouraging signs on the political horizon. The Clinton Administration is far less reluctant than its predecessors to embrace measures that can be construed as industrial policies. The Clinton economic plan includes substantial federal investments to help stimulate the civilian economy and address the defense transition problem. Its programs target transportation, the environment, advanced manufacturing, information infrastructure, education, civil aviation, and economic conversion, among others.

Congress also appears more supportive of civilian technology and economic adjustment and conversion programs, especially in the new, more receptive political environment. Last year, the Senate Task Force on Defense Transition, chaired by Senator David Pryor (D-Arkansas), made a number of far-sighted recommendations, including economic adjustment and conversion measures to reduce the impact of defense cuts, increased support for civilian industrial technology (through R&D tax credits and targeted government spending), and the creation of a National Environmental Technologies Agency to fund environmental R&D.[97]

In its FY 1993 defense authorization and appropriations bills, Congress provided $1.7 billion for a defense reinvestment and conversion package largely drawn form the Pryor Task Force proposal. In the 103rd Congress, both the Senate and House have introduced major bills that target civilian technology needs that have a greater chance of being passed than they would have in earlier years.[98] While all these initiatives from both the White House and Capitol Hill have serious shortcomings they represent significant steps in the right direction.

There are also positive signs at the grassroots level. In recent years, community, labor, and public interest groups have sought to influence industrial and S&T policies. These new players include the National Toxics Coalition, the Council for Responsible Genetics, the Federation for Industrial Renewal, the National Commission on Economic Conversion and Disarmament, and state and local conversion projects in Minnesota, Maine, Washington, and California. Labor unions also have begun to articulate their own concerns and agendas.

A pioneering national grassroots effort to inject social criteria and citizen representation into S&T policy is the Campaign for Responsible Technology (CRT), which brings together community, environmental, and labor groups concerned with environmental and occupational health issues in high-tech manufacturing. The campaign is trying to get the government-subsidized Sematech consortium of semiconductor manufacturers to broaden its mandate to include environmental, labor, and community concerns.[99] CRT registered a small but significant success in the spring of

1992, when the House Armed Services Committee followed a CRT recommendation to include $10 million in the 1993 Defense Authorization bill earmarked for Sematech to develop manufacturing methods that are environmentally safe.

Working closely with CRT is the 21st Century Project, sponsored by Computer Professionals for Social Responsibility (CPSR), the nation's only public interest organization of information technology professionals. CPSR aims to democratize technology investment decisions and to integrate a national perspective with the work of citizens at the local level. Its national presence will be developed through ongoing research and a sustained effort of public outreach.[100]

The 21st Century Project's research program seeks to provide the intellectual content for a new S&T agenda, which will be based on the findings of four working groups: democratic management of S&T, sustainable development, quality of work in the information age, and computers and communication infrastructure. Participants include public interest activists, labor leaders, academics, policy experts, and technical professionals. The study will present a model R&D program built around a set of key societal problems, together with potentially relevant applications of information technology.

It is too early to predict the outcomes of these efforts. One thing is certain, though. The nation can no longer afford to pursue a do-nothing *laissez-faire* industrial strategy. Nor can it allow narrow military and commercial interests to dictate its economic future. The public is more than ready for new S&T policies to build a society that is healthy, economically stable, and environmentally sustainable. Whatever its flaws, the Clinton Administration has put foward a vision of economic change upon which we can now build. All that's needed is the political will.

Notes

1. Since 1984, when President Reagan created the President's Commission on Industrial Competitiveness headed by Hewlett-Packard CEO John Young, a number of government organizations and private sector commissions, such as the Council of Competitiveness, the Cuomo Commission on Trade and Competitiveness , the MIT Commission on Industrial Competitiveness, the Carnegie Commission on Science, Technology and Government, the National Academy of Engineering, the National Advisory Committee on Semiconductors, the Department of Commerce, and the Office of Technology Assessment, among others, have issued findings and recommendations that are highly consistent. See, for example, Council on Competitiveness, *Picking Up the Pace:*

The Commercial Challenge of American Innovation (Washington, DC: 1988); The Cuomo Commission on Trade and Competitiveness, Lee Smith, Director and Editor, Lewis B. Kaden, Chairman, *The Cuomo Commission Report, A New American Formula for a Strong Economy* (New York: Simon & Schuster, Inc., 1988); Michael L. Dertouzos, Richard K. Lester, Robert M. Solow, and the MIT Commission on Industrial Competitiveness, *Made in America: Regaining the Productive Edge* (Cambridge, MA: The MIT Press, 1989).

2. John D. Steinbrunner, "The Changing Nature of Military Threat," in Susan L. Sauer ed., *Science and Technology and the Changing World Order*, Colloquium Proceedings, 12-13 April 1990 (Washington, DC: American Association for the Advancement of Science, Committee on Science Engineering and Public Policy, 1990), pp. 109-114.
3. Steven Greenhouse, "Hardly Laissez-Faire," *The New York Times*, 27 June 1992, p. 9.
4. These budget figures are only for basic and applied research and development, and do not include government totals for R&D facilities. Colin Norman, "Science Budget: Selective Growth," *Science*, 255:5045 (7 February 1992), pp. 672-75.
5. Remarks of Governor Bill Clinton, Wharton School of Business, University of Pennsylvania, Philadelphia, PA, 16 April 1992; Alan Murray, "Democratic Frontrunner Backs Industrial Policy with a Populist Twist," *The Wall Street Journal*, 23 April 1992, pp. A1, A6.
6. These figures may be understated as they do not include military-related portions of the NASA R&D budget that are not officially included in the defense budget function.
7. For a detailed elaboration of the "closet industrial" policy argument see Ann Markusen and Joel Yudken, *Dismantling the Cold War Economy* (New York: Basic Books, 1992), especially Chapters 3-5.
8. Aerospace, for example, registered nearly $37 billion in exports in 1990, contributing $26 billion to the U.S. trade balance. Aerospace Industries Association of America, "1990 Year-End Review and Forecast: An Analysis," mimeo (Washington, DC: 12 December 1990), Table VI.
9. For discussion of the S&E work force and the military-industrial sector, see Joel Yudken and Ann Markusen, "The Labor Economics of Conversion: Prospects for Military-Dependent Engineers and Scientists," in Martha Gilliland and Patricia MacCorquodale, eds., *Engineers and Economic Conversion: From the Military to the Marketplace* (Berlin/Heidelberg: Springer-Verlag, in press), Table 7.
10. Source of data is unpublished from the National Science Foundation's Survey of Scientists and Engineers (SSE), compiled by Ann Markusen. Yudken and Markusen, "Labor Economics of Conversion," Table 8.
11. For further elaboration see Markusen and Yudken, *Dismantling*, Chapter 3.
12. Ann Markusen, *Steel and Southeast Chicago: Reasons and Remedies for Industrial Renewal*, Report to the Mayor's Task Force on Steel and Southeast Chicago, (Evanston, IL: Center for Urban Affairs and Policy Research, Northwestern University, 1985).

13. Steven Greenhouse, "There's No Stopping Europe's Airbus Now," *New York Times*, 23 June 1991, pp.1, 6.
14. "Milspec" refers to specifications for parts and items formally detailed in the body of "Military Specifications and Standards" published by the Department of Defense.
15. Defense Science Board, *Report of the Defense Science Board Task Force on Defense Semiconductor Dependency*, prepared for the Office of the Under Secretary of Defense for Acquisition (Washington, DC: February 1987).
16. This was reported recently by the General Accounting Office. Cited in Jacques S. Gansler, "Restructuring the Defense Industrial Base," *Issues in Science and Technology* (Spring 1992), pp. 50-58. A classic illustration of how a declining domestic manufacturing base is troubling the DoD, concerns Avtex, a manufacturer or rayon fibers for the apparel industry. In November 1988, Avtex announced that it was closing its doors due to foreign competition. The Office of Technology Assessment writes that this move "sent shock-waves through the DoD and NASA when it was discovered that Avtex was the only producer of fibers that were critical to the production of missiles and rockets... Negotiations were soon completed to keep Avtex open." U. S. Congress, Office of Technology Assessment (OTA), *Holding the Edge: Maintaining the Defense Technology Base*, OTA-ISC-420 (Washington, DC: U.S. Government Printing Office, April 1989), p. 33.
17. OTA, *Holding the Edge*, Chapter 9, pp.161-187; and Gansler, "Restructuring the Defense Industrial Base," p. 53.
18. For more in-depth discussion of these issues see Jay Stowsky, "Beating Plowshares into Double-Edged Swords: The Impact of Pentagon Policies on the Commercialization of Advanced Technologies" (Berkeley, CA: Berkeley Roundtable on the International Economy, April 1986).
19. Gansler, "Restructuring the Defense Industrial Base," p. 52. Gansler estimates that there are about 27,000 military specifications and 7,000 unique military standards.
20. OTA, *Holding the Edge*, p. 162.
21. For discussion of "spin-on" versus "spin-off" see Michael Borrus and John Zysman, "The Highest Stakes: Industrial Competitiveness and National Security," BRIE Working Paper 39 (Berkeley, CA: Berkeley Roundtable on the International Economy, April 1991), especially p. 22ff.
22. Gansler, "Restructuring the Defense Industrial Base;" OTA, *Holding the Edge*, especially Chapter 9, pp. 161-187.
23. U.S. Department of Defense, *Critical Technologies Plan* (for the House Armed Services Committee, United States Congress) (Washington, DC: 15 March 1990).
24. "Critical," "emerging," and "enabling" technologies are terms often used interchangeably. "Critical technologies" are meant to refer to technologies that will lead to new products and industries on a large scale or produce large advances in productivity and quality. They are also thought to be important because they will drive the next generation of technological development. The Commerce Department defines "emerging technologies" as those with potential to ad-

vance "productivity and quality," and create a "multitude of new products and services."

25. Gansler, "Restructuring the Defense Industrial Base," p. 57.
26. DARPA officials in the late 1980s have even proudly equated the role of their agency to that of Japan's Ministry of International Trade and Industry (MITI).
27. Eliot Marshall, "Beating Swords into . . . Chips?" (box in "U.S. Technology Strategy Emerges") *Science*, 252:5002 (5 April 1991), p. 22; and William J. Broad, "Pentagon Wizards of the Technology Eye Wider Civilian Role," *New York Times*, 22 October 1991, pp. C1, C11.
28. Branscomb, "Toward a U.S. Technology Policy," p. 51.
29. OSTP director Bromley plans to establish other FCCSET committees over the next couple of years. Joseph Palca, "It Ain't Broke, But Why Not FCCSET?" *Science*, 251:4995 (15 February 1991), pp. 736-37.
30. Marshall, "Industrial R&D Wins Political Favor," *Science*, 250:5148 (3 March 1992) pp. 1502-1509.
31. Eliot Marshall, "Headed for a Brick Wall," *Science* 256:5056 (24 April 1992), p. 439.
32. The report also proposes changes in the roles of the National Security Council and other Executive offices to make technology policy a central concern of both national security and economic policy at the highest levels of the executive branch. See Carnegie Commission on Science, Technology and Government, *Technology and Economic Performance: Organizing the Executive Branch for a Stronger National Technology Base* (New York: CCST&G, September 1991).
33. Colin Norman, "The Academy Gives a Hard Push," *Science*, 256:5053 (3 April 1992), p. 23.
34. The FY 1993 bills provide budget authority of $845 million for Department of Defense Dual-Use Technology Reinvestment Programs. A few other dual-use or related programs supported by DARPA, such as the Sematech consortium, which received $100 million in the FY 1993 defense bills, are not included in this figure.
35. See Richard R. Nelson and Sidney G. Winter, *An Evolutionary Theory of Economic Change* (Cambridge, MA: Belnap Press of Harvard University Press, 1982); Richard R. Nelson, *High-Technology Policies, A Five Nation Comparison*, (Washington, DC: American Enterprise Institute for Public Policy Research, 1984); Giovanni Dosi, Christopher Freeman, Richard Nelson, Gerald Silverberg, and Luc Soete eds., *Technical Change and Economic Theory* (London: Pinter Publishers, 1988); David C. Mowery and Nathan Rosenberg, *Technology and the Pursuit of Economic Growth* (New York: Cambridge University Press, 1991).
36. Although few of the "dual-use" proposals directly refer to Schumpeterian and neo-Schumpeterian analyses, more than one individual in this camp has referred to himself as a self-avowed neo-Schumpeterian in private conversations with the author. The works of Nelson, Dosi, Rosenberg, *et al.* also are frequently cited in the growing body of published literature examining government policy and high-tech industrial issues.
37. Christopher Freeman, "Introduction," in Giovanni Dosi *et al.*, *Technical Change and Economic Theory*, pp. 1-8.

38. For example, Nelson and Winter contend that treating innovation "within an evolutionary model provides a far better basis for modeling economic growth fueled by technical advance than does the neoclassical model," even if the latter is modified to include variables representing technical change. In their work, they "develop the point that an evolutionary theory of growth offers a framework that is far more capable of integrating micro and macro aspects of technical advance than is the more orthodox, formal approach." Nelson and Winter, *An Evolutionary Theory of Economic Change*, p. 22
39. Nathan Rosenberg, *Perspectives on Technology* (New York: Cambridge University Press, 1976), p. 123.
40. These terms were first used in Rosenberg, *Perspectives.* Dosi discusses these terms in "The Nature of the Innovative Process," in Giovanni Dosi *et al., Technical Change and Economic Theory*, pp. 221-238, 225ff.
41. Nelson, *High-Technology Policies*, p.11.
42. See Mowery and Rosenberg, *Technology and the Pursuit of Economic Growth* , Chapter 1, for a critical assessment of private sector difficulties in appropriating basic research results.
43. Some high-tech proposals for industrial consortia, however, such as for high-definition TV, have called for product prototyping and development. "Super Television, The High Promise—and High Risks—of High Definition TV," *Business Week* (30 January 1989), pp. 56-66.
44. The term industrial policy has had a controversial history. It means many things to different people, but it can be generally defined as an "attempt by the government to guide or shape industrial trends through the use of tax incentives, subsidies and loans." Kim Moody, "Industrial Policy," *Labor Notes* (27 July 1983). See note 55 below.
45. Kenneth Flamm, *Creating the Computer* (Washington, DC: The Brookings Institution, 1987), pp. 68-75; and *Targeting the Computer* (Washington, DC: The Brookings Institution, 1987), p.44ff.
46. Nelson, *High-Technology Policies*, p.13.
47. For a comparative review of these industries up through the mid-1980s see *ibid*.
48. ESPRIT, the acronym the European Strategic Programme of Research in Information Technology, is funded by EEC funds; RACE is the acronym for R&D in Advanced Communications for Europe and BRITE is Basic Research in Industrial Technologies for Europe. Having a broader scope are the European Eureka Program, with applications targeted to commercial products, manufacturing, transportation, health, environmental problems, and other areas, and the Japanese Human Frontier Science Program, which focuses on human biological functions and their artificial applications, with the goal of stimulating advances in biotechnologies, diagnostics, instrumentations, and pharmaceuticals. Mario Pianta, "High Technology Programmes: For the Military or for the Economy," *Bulletin of Peace Proposals*, 19:1 (1988), pp. 53-79.
49. Daniel S. Greenberg, *The Politics of Pure Science* (New York: New American Library, 1967), p. 58; Markusen and Yudken, *Dismantling*, pp. 40, 103.
50. Mowery and Rosenberg, *Technology and the Pursuit of Economic Growth*, p.184.
51. Linda R. Cohen and Roger G. Noll, *The Technology Pork Barrel* (Washington, DC: The Brookings Institution, 1991), pp. 1-2.

52. In what he aptly terms the "medical technology 'arms race'," *New York Times* reporter Andrew Pollack reports "a growing recognition that the uncontrolled use of high-technology medical equipment and procedures. . . helps drive the relentless increase in medical cost." Pollack observes that medical practitioners are rewarded more for using sophisticated treatments than for routine health care. Hence, there is a tendency to overuse advanced, expensive medical procedures. Andrew Pollack, "Medical Technology 'Arms Race' Adds Billions to the Nation's Bills," *New York Times*, 29 April 1991, pp. A1, B8.
53. Markusen and Yudken, *Dismantling,* especially Chapter 3.
54. This includes DoE spending on defense activities, but does not include the $5-6 billion spent on military construction. These figures hold true for most of the 1980s, though projected budgets into the early 1990s suggests a sizeable drop in procurement spending, down to around 20 percent, though R&D is expected to remain a stable 12-13 percent of the defense budget. Steve Kosiak and Paul Taubl, *Analysis of the Fiscal Year 1993 Defense Budget Request* (Washington, DC: Defense Budget Project, 11 March 1992), Table 7.
55. In actuality, these industries are composites of several related industries. See also Markusen and Yudken, *Dismantling,* pp. 35, 57. Features that constitute an industrial policy include: the targeting of sectors as growth leaders; substantial incentives and subsidies for R&D; provision of capital for plant, equipment, and operating expenses; encouragement for various forms of industrial collaboration and planning; the monitoring and shaping of competition, such as through selective bailouts and contracting; government guaranteed markets for industry outputs, especially in the early stages of development; export supports and trade protection; adjustment assistance for firms, workers, and communities affected by facility closures. Every one of these practices has been employed by the U.S. government in its military industrial policy. Ann Markusen, "Defense Spending: A Successful Industrial Policy?" *International Journal of Urban and Regional Research,* 10:1 (1986), pp. 105-22.
56. Borrus and Zysman, "The Highest Stakes," p. 27.
57. Giovanni Dosi, *Technical Change and Industrial Transformation* (New York: St. Martin's Press, 1984), p.18.
58. For an earlier critical elaboration of the weaknesses of the high-tech "dual-use" industrial policy, see Yudken and Black, "Targeting National Needs."
59. Some ex-officials have privately noted that the visionary leadership in DARPA's heyday during the sixties and seventies, notably in its computing programs, that was responsible for the agency's cutting-edge reputation, is no longer there. More narrowly focused, military-oriented personnel, more concerned about short-term "deliverables" over long-term advancement, are now said to dominate.
60. Yudken and Black, "Targeting National Needs," p. 270, n. 65, for an explanation of how this figure was derived.
61. Mowery and Rosenberg, *Technology and the Pursuit of Economic Growth,* p. 144.
62. In Joel Yudken, "Report and Retrospective on the Workshop on Technology, Industrial Policy and National Needs." Sponsored by the Project on Regional and Industrial Economics (PRIE), Rutgers University, 8-9 April 1991 (New Brunswick, NJ: PRIE, March 1992). The Coalition on New Office Technology is

a coalition of 45 Massachusetts based organizations, including large office worker and service sector unions, concerned with technology impacts in the workplace, especially issues such as electronic monitoring and health hazards.

63. Cited in Dwight B. Davis, "Assessing the Strategic Computing Initiative,"*High Technology* (April 1985), pp. 41-49.
64. For example, "social constructivists," such as Bruno Latour and Michel Callon, perceive scientific research and its products as the result of social interactions, networks, and "enrolling" activities among researchers. One of the best known works of this genre is Bruno Latour, *Science in Action* (Cambridge, MA: Harvard U. Press, 1987).
65. I owe this definition to Charley Richardson, director of the Technology and Work Program, University of Massachusetts-Lowell, who used it in a meeting of the Campaign for Responsibility in Washington, DC, 2 February 1992.
66. Susan Walsh Sanderson, *The Consumer Electronics Industry and the Future of American Manufacturing* (Washington, DC: Economic Policy Institute, 1989), p. 13.
67. Scott Sullivan, "The New Supertrains," *Newsweek* 31 July 1989, pp. 46-48; Marcus Mabray, "Why the Fuss? Just Ask the Japanese,"*Newsweek* 31 July 1989, p. 48.
68. U.S. Congress, Office of Technology Assessment, *Commericializing High-Temperature Superconductivity*, OTA-ITE-388 (Washington, DC: U.S. Government Printing Office, June 1988), p. 74.
69. Tom Forester ed., *The Materials Revolution* (Cambridge, MA: The MIT Press, 1988), p.12.
70. Malcolm Gladwell, "The Changing Face of Science: Bernadine Healy Shakes the Old Guard," *Washington Post National Weekly Edition*, 29 June-5 July 1992, pp. 6-8. The *Post* reports that Healy "comes from a profession, cardiology, that decided to explore heart disease risk factors by studying 15,000 men and zero women; that looked at aspirin as a preventative therapy for coronary disease in 22,000 and zero women; and that tried to answer the question of whether estrogen was protective against heart disease in women by conducting a study of the role of estrogen in preventing heart disease in men."
71. Ken Geiser, "The Greening of America, Making the Transition to a Sustainable Economy" *Technology Review*, 94:6 (August/September 1991), pp. 64-72.
72. Markusen and Yudken, *Dismantling*, especially Chapter 9.
73. *Ibid.*
74. Paul A. Krugman, "Myths and Realities of U.S. Competitiveness," *Science* 254:5033, 8 November 1991, pp. 811-15.
75. Jeff Faux and Todd Schafer, "Increasing Public Investment, New Budget Priorities for Economic Growth in the Post-Cold War World" (Washington, DC: Economic Policy Institute, October 1990).
76. For discussion of adjustment problems, see Markusen and Yudken, *Dismantling*, Chapters 6-8; and U.S. Congress, Office of Technology Assessment, *After the Cold War: Living With Lower Defense Spending* OTA-ITE-524 (Washington, DC: U.S. Government Printing Office, February 1992).
77. For elaboration of this argument, see Yudken, "Technology, Defense Conversion and Economic Development."

78. Office of Science and Technology Policy, *Report of the National Critical Technologies Panel* (Washington, DC: U.S. Government Printing Office, 1991), p. 122.
79. Geiser, "The Greening of America."
80. William K. Stevens, "2 Percent of G.N.P. Spent by U.S. on Cleanup," *The New York Times*, 23 December 1990; Valjean McLenighan, *Sustainable Manufacturing, Saving Jobs, Saving the Environment,* special issue of *The Neighborhood Works,* (Chicago, IL: Center for Neighborhood Technology, 1990), p.13.
81. Michael Schrage, "Japan Is Set to Meet America on the Environmental Products Battleground," *The Washington Post,* 27 March 1992, p. F3. The $300 billion estimate is from an Office of Technology Assessment Project Description sheet, "American Industry and the Environment: Implications for Trade and U.S. Competitiveness," unpublished (25 July 1991).
82. Computer Professionals for Social Responsibility has been leading the effort in the U.S. Aki Namioka and Douglas Shuler eds., "PDC'90, Proceedings of the Participatory Design Conference," Seattle, WA, 31 March-1 April 1990 (Palo Alto, CA: Computer Professionals for Social Responsibility Workplace Project).
83. Moshe Ben-Akiva, David Bernstein, Anthony Hotz, Haris Koutsopoulus, and Joseph Sussman, "The Case for Smart Highways," *Technology Review*, 95:5 (July 1992), pp. 38-47.
84. Jonathan P. Hicks, "A Faster Path to Finished Steel," *New York Times*, 7 April 1991, p. D9.
85. For example, Hughes Aircraft Company recently developed a solder flux that significantly reduces the use of chlorofluorocarbon (CFC)-based solvents in electronics manufacturing, which, at the same time, reportedly speeds up manufacturing and enhances productivity. B.D. Nordwall, *Aviation Week and Science Technology*, 17 February 1992, p. 45.
86. James C. Peterson, "Citizen Participation in Science Policy," in James C. Peterson ed., *Citizen Participation in Science Policy* (Amherst, MA: The University of Massachusetts Press, 1984), pp. 1-17.
87. Peterson ed., *Citizen Participation in Science Policy*, p. 10
88. Dorothy Nelkin, "Science and Technology Process and the Democratic Process," in Peterson, *Citizen Participation*, pp. 18-39.
89. Rachelle Hollander, "Institutionalizing Public Service Science: Its Peril and Promise," in Peterson, *Citizen Participation*, pp. 75-95.
90. These are technical assistance centers which would foster fruitful interactions between academic research laboratories and local businesses, labor, and communities. Philip Shapira, *Modernizing Manufacturing, New Policies to Build Industrial Extension Services* (Washington, DC: Economic Policy Institute, 1990).
91. This center was set up involving labor and community participation to help the local textile industry be more productive, but ran into trouble when it tried to institute expensive, state-of-the-art technologies. It turned out that technical assistance to aid day-to-day technical problems was much more useful. Richard Kazis, "Rags to Riches? One Industry's Strategy for Improving Productivity," *Technology Review*, 92:6 (August 1989), pp. 42-53.
92. The IAM also set up a technology center outside Washington, DC, for training members to handle technology issues, and has pushed for inclusion in regional NIST-sponsored manufacturing technology projects. Interview with Richard

Greenwood, former Special Assistant to retired IAM President William Winipisinger, 31 January 1989.

93. OTA, *Commercializing*, p. 37.

94. Doron P. Levin, "Saturn: An Outpost of Change in G.M.'s Steadfast University," *New York Times*, 17 March 1991.

95. Gwendolyn Freyd, "Putting Engineering in Context," MIT Reporter, *Technology Review*, 95:2 (February/March 1992), pp. 12-13.

96. Women, African-Americans, and Latinos made up 16 percent, 3 percent, and under 2 percent, respectively, of the total S&E work force in 1988. The 1988 figures are estimates, but are consistent with earlier year actual figures. These groups do much better in the sciences than in engineering. Women make up 29 percent of the scientists, but only 5 percent of the engineers. One reason is that the science occupations include the life science and psychology fields which traditionally have attracted and allowed many more women to participate than the physical sciences or engineering. African-Americans made up 4 percent of the scientists, and 2 percent of the engineers. National Science Foundation, *Science and Engineering Personnel: A National Overview. Special Report*, NSF 90-310 (Washington DC, 1990). See also Yudken and Markusen, "Labor Economics of Conversion," Table 2.

97. It also would require the DoD to set aside R&D funds for "dual-use" technologies relating to environmental cleanup, energy efficiency, transportation, computer and communications technology, and others. Senate Task Force on Defense Transition, "Background on Recommendations" (Washington, DC, 21 May 1992), unpublished.

98. The Senate bill entitled the "National Competitiveness Act of 1993," introduced by Senator Ernest F. Hollings (D-South Carolina), chair of the Senate Committee on Commerce, Science, and Transportation, greatly expands the civilian technology programs of the Department of Commerce.

99. Lenny Siegel, Ted Smith, and Rand Wilson, "Sematech, Toxics, and U.S. Industrial Policy: Why We Are Concerned" (1990); Rand Wilson, "From Day Care to DARPA: Bargaining for A New Industrial Policy" (Somerville, MA: Campaign for Responsible Technolgy, 1991).

100. Michael Alexander, "Advocacy Group Pushes for Shift in R&D Funding," *Computerworld*, 11 November 1991.

Chapter 2

Technological Politics As If Democracy Really Mattered: Choices Confronting Progressives

by Richard Sclove

A century and a half ago Alexis de Tocqueville described a politically exuberant United States in which steaming locomotives could not restrain citizens' enthusiasm to involve themselves in politics and community life:

> In some countries the inhabitants seem unwilling to avail themselves of the political privileges which the law gives them; it would seem that they set too high a value upon their time to spend it on the interests of the community; and they shut themselves up in a narrow selfishness. . . . But if an American were condemned to confine his activity to his own affairs, he would be robbed of one half of his existence; he would feel an immense void in the life which he is accustomed to lead, and his wretchedness would be unbearable.[1]

That is not today's United States, in which a bare majority of eligible voters participate in presidential elections while usually even fewer engage in local politics.[2] The causes of Americans' political disengagement are complex, but one culprit, more significant and intricate than commonly believed, is technology. Consider an instructive story from across the Atlantic.

During the early 1970s running water was installed in the houses of Ibieca, a small village in northeast Spain. With pipes running directly to their homes, Ibiecans no longer had to fetch water from the village fountain.

Families gradually purchased washing machines, and women stopped gathering to scrub laundry by hand at the village washbasin. Arduous tasks were rendered technologically superfluous, but village social life was unexpectedly altered. The public fountain and washbasin, once scenes of vigorous social interaction, became nearly deserted. Men began losing their sense of familiarity with the children and donkeys that once helped them haul water. Women stopped gathering at the washbasin to intermix scrubbing with politically empowering gossip about men and village life. In hindsight the installation of running water helped break down the Ibiecans' strong bonds—with one another, with their animals, and with the land—that had knit them together as a community.[3] Painful in itself, such loss of community carries a specific political cost as well: as social ties weaken, so does a people's capacity to mobilize for political action.[4]

Is this a parable for our time? Like Ibiecans, we acquiesce in seemingly benign or innocuous technological changes. Ibiecans opted for technological innovations promising convenience, productivity, and economic growth. But they did not anticipate the hidden costs: greater inequality, social alienation, and steps toward community disintegration and political disempowerment. Does technological change invariably embody a Faustian trade-off between economic reward and sociopolitical malaise? No, not invariably. But the best hope for escaping such trade-offs is to develop a full-blown democratic politics of technology—something that even political progressives have not begun to conceive.

Economic Performance vs. Democracy

In recent years progressive thinkers generally have agreed with conservatives that technology policy should be aimed at conventional economic objectives: productivity, growth, and international competitiveness. But where conservative rhetoric has favored *laissez-faire* means to achieve these goals, proposed progressive policies have promoted greater government intervention, such as new civilian authorities to guide technological development away from militarism and toward basic societal needs like health care, environmental protection, infrastructure improvements, and stronger industries.[5] The more daring of these proposals are avowedly populist, seeking to create decentralized institutions for grassroots participation in scientific and technological decisionmaking.[6] Nevertheless, because progressives would orient technology policy fundamentally toward economic objectives, their disagreement with conservatives is over means more than ends.

Many of these progressive policies make considerable sense and, with the election of Bill Clinton, some seem poised to fly politically. They are also unquestionably better than the hodge-podge of military-industrial strategies that have dominated technological evolution in the United States since World War II and that appear increasingly anachronistic now that the Cold War is over. Nevertheless, progressive strategies proposed to date consistently fail to come to terms with the *latent* role of technologies in shaping social and political relations.[7] Contemporary technologies contribute indirectly to a variety of social ills and hinder participatory decisionmaking about technology and in general. Supporters of existing structures of technological decisionmaking, as well as advocates of change, have yet to grasp this point.

Of all the social impacts of technology, perhaps the most worrisome are the adverse effects on democracy. A vigorous democracy is a necessary tool to address all other areas of social concern.[8] If the social and political potency of technology is not taken into account, the best we can hope for is a variant of Ibieca's fate: improvements in productivity or marginal steps toward meeting social needs that are nonetheless associated with political disengagement, attenuated community bonds, experiential divorce from nature, individual purposelessness, and expanding disparities in wealth.[9] Is that what we dream and sweat for?

Take a seemingly innocuous convenience like the microwave oven. How often does the ease with which we now can prepare food ironically result in households rarely finding the time to sit down and eat together?[10] Or consider how air conditioning and television have helped subvert the custom of neighbors gathering on outdoor stoops and porches during the summertime. Neighbors who hardly know one another may experience more than a sense of loss or loneliness; they are also apt to lack the social and organizational structures needed to identify, analyze, and address common concerns politically.[11] Thus, while often no one technological innovation may be of great social consequence, the cumulative effect of such innovations can be profound. Even infrastructural technologies that we take for granted, such as urban sewage systems (discussed shortly), often are designed in ways that prove detrimental to democracy.

Fortunately, it is possible to envision alternative technological strategies and designs that not only can fulfill vital economic and social needs but also can help sustain democratic community, civic engagement, and social justice. Thus the goal is not to reject all technology outright—clearly a ludicrous proposition—but rather to become more discriminating in how we design, choose, and use technologies. Moving in this direction would require our creating democratic institutions for evolving democratic technologies. It would also require our taking a significant political risk—to grant equal or greater priority to democracy over short-run economic objectives.

Those who champion a technology policy based on social needs and who emphasize improved economic productivity and growth are on politically safer ground. But to date there is no evidence that proponents of this strategy have considered the risk that a more "competitive" technological order could create a world with greater inequality, passivity, and alienation. In contrast, the decision to espouse a democratic politics of technology, while confronting serious short-run political obstacles, embodies the chance to create a world of expanding personal and collective empowerment, a world in which concocting and consuming "New Improved!" deodorants, hi-tech designer sneakers, or chip-laden electronic widgetry need no longer preempt the quest to revitalize democracy, to address real social needs, and to fulfill other human aspirations.[12]

This does not mean that technology is necessarily the most important factor influencing political life. But technology is sufficiently important—and so inextricably intertwined with other factors like the distribution of wealth, race and gender relations, international affairs, and so on—that we must learn to subject it to the same rigorous political scrutiny that we apply to the other factors.

But can the proposition that democracy should take precedence over economics really be taken seriously? Absolutely. After all, was not one lesson of Eastern Europe's anti-Communist revolutions in 1989 that an economy unsupervised by democracy is bad not only because it *is* undemocratic (bad as that may be) but also because it risks injustice, ecological spoliation, and gross economic inefficiency? Or would it be reasonable to have required the Founding Fathers to subject each article of the proposed Bill of Rights to an economic cost-benefit test? Would we be satisfied if they had rejected protection of free speech on the grounds that the supposed benefits could not hold a candle against the obvious economic costs?

Technologies do not merely deliver sundry consumer benefits (as well as sundry hazards and irritations); they also comprise part of a society's core political infrastructure. Technologies establish an intricate, pervasive, and yet remarkably under-appreciated network of politically consequential social influences, opportunities, constraints, and inducements. Notwithstanding the startling degree to which we remain oblivious to it, this network is as significant as the more explicit network of influences, opportunities, constraints, and inducements established by a society's legal and tax codes.

Technological democratization—that is, the establishment of more democratic processes to produce a more democratic technological order—is vital for several basic reasons. First, it is a crucial (but almost entirely neglected) requisite for advancing the democratization of society in general. Second, as shown below, it directly addresses a number of commonly perceived social problems, such as the decline of community and the degraded nature of work. Third, by contributing to the overall democratization of society, technological democratization would empower citizens to begin addressing the other social consequences of technology as well as other societal problems not attributable to technology.

Technology as a Social Structure

Technologies are more than mere tools to accomplish narrowly defined objectives. Technologies also function as social structures. Other familiar social structures include laws, dominant political and economic institutions (such as legislatures, courts, and corporations), and systems of cultural beliefs. All of these structures are social creations that profoundly influence one another's evolution, as well as the course of history and texture of daily life. Social structures—considered collectively—shape society's fundamental political relationships and processes. To appreciate the subtle means by which technologies can exert structural influence, consider Ibieca once more.

First, upsetting Ibieca's traditional pattern of water use undermined ways in which the village perpetuated itself as a self-conscious community. Technologies helped to re-structure social relations. But notice that technologies tend to do this *independently of their nominally intended, or "focal," purposes.* We do not normally regard fountains and pipes (or for that matter microwave ovens, hypodermic syringes, garden hoses, or numerically controlled machine tools) as devices that shape patterns of human relationship, but that nevertheless is one of their pervasive latent, or "non-focal," tendencies.[13]

Second, clusters of focally unrelated technologies often interact non-focally to produce structural results that no one technology could produce alone. In Ibieca the introduction of water pipes gave residents—who formerly hauled water by donkey—an added incentive to replace donkeys with tractors in field work. (The fewer tasks a donkey is asked to perform, the less economical it is to maintain it). The elimination of donkeys increased the villagers' dependence on outside jobs for the cash needed to finance and operate their new tractors and washing machines. Thus, for social insight it's not enough to consider just one kind of technology at a time (say, water pipes alone); all the different artifacts, practices, and systems that comprise a society's technological order (e.g., water pipes, washing machines, tractors, etc.) must be analyzed.

These generalizations have important implications for progressive technological agendas. For instance, the fact that a technology typically produces its most profound social effects indirectly—often in concert with other, seemingly unrelated technologies—suggests the problem with technological strategies preoccupied with *focally* addressing social needs. Of course, we should develop new technologies that address such needs. But we should do so while ensuring that the combined focal *and* non-focal impacts of the entire technological order contribute constructively to democracy, grassroots empowerment, and other social objectives. If Ibiecans want new ways to manage their water use or to plow their fields, can they craft means that will not sacrifice other desirable aspects of community life? (Perhaps a centrally located laundry cooperative would be better for the village than household washing machines.) Similarly, if Americans are prepared to begin substituting better health care or better public transportation for their self-destructive addiction to advanced military hardware, can we adopt approaches that will not inadvertently reinforce national social malignancies, such as the decay of inner cities and the rise of atomistic suburbs?[14]

The preceding arguments also highlight the inadequacy of becoming too preoccupied either with sexy, cutting edge technologies, such as biotechnologies, semiconductors, or cryogenics, or with technologies that are *focally* political, such as computerized voting and interactive television.[15] While it is foolish to overlook such technologies, it is equally wrong to ignore the vast majority of more familiar or focally innocuous technologies—pipes, washing machines, air conditioners, electricity distribution networks, and so on—that have significant non-focal impacts.

Technology and Democracy

The approach to technology policy proposed here is grounded morally in the belief that people should be able to shape the basic social circumstances of their lives. It is aimed at organizing society along relatively equal and participatory lines, at achieving a system of egalitarian decentralization and confederation that Rutgers political scientist Benjamin Barber calls "strong democracy."[16] Historic examples of strong democracy include New England town meetings, the confederation of self-governing Swiss villages and cantons, and the tradition of trial by a jury of peers. Strong democracy also is apparent in the methods or aspirations of various social movements, such as the late nineteenth century American Farmers Alliance, the 1960s civil rights movement, and the 1980s uprising of Solidarity in Poland.[17] In each of these cases ordinary people claimed the rights and responsibilities of active citizenship.

If citizens ought to be empowered to participate in determining their society's basic structure and if technologies *are* an important part of that structure, it follows that technological design and practice should be democratized. Substantively, technologies must be compatible with our fundamental interest in strong democracy. And procedurally, people from all walks of life must have expanding opportunities to shape the evolving technological order.

Design Criteria for Democratic Technologies

Table 1 presents some criteria for distinguishing among technologies based on their compatibility with democracy. The criteria are labeled "provisional" because they are neither complete nor definitive. Rather, they are intended to provoke political debate that can lead to an improved set of criteria.

Each criterion is intended to fulfill the institutional requirements for strong democracy: democratic community, democratic work, or democratic politics.[18] Technological decisions should attend initially and foremost to strengthening democracy, because democracy provides the necessary circumstances for deciding freely and fairly what other considerations must be taken into account in technological (and non-technological) decisionmaking. Until we do this, technologies will continue to hinder the advancement of other social objectives in subtle yet significant ways.

A series of examples can help explain these criteria and the feasibility of designing technologies that can satisfy them. Before proceeding, however, one clarifying note is in order. Each of the following examples illustrates a worthy social and democratic goal in its own right. However, isolated technological changes of this sort cannot be expected to represent a significant improvement in the overall democratization of society. The latter result will require multiple democratic design criteria, applied simultaneously to diverse technologies by citizens who employ broadly democratized processes of technological decisionmaking. In other words, all the elements of a complete democratic politics of technology should converge at one time.

Criterion A: Technology and Democratic Community

Egalitarian community life is important to strong democracy because it enhances citizens' mutual respect, shared understanding, political equality, and social commitment. It empowers individuals within collectivities to challenge unjust concentrations of power. Unfortunately, diverse technological developments have contributed to the decline of community. The noise and danger of automobile traffic, detached single-family homes, air conditioning, and television all have isolated families away from one another and undermined a sense of collective purpose. This has been

Table 1
A PROVISIONAL SYSTEM OF DESIGN CRITERIA FOR DEMOCRATIC TECHNOLOGIES

Toward DEMOCRATIC COMMUNITY:

A. Seek a balance among communitarian/cooperative, individualized, and intercommunity technologies. Avoid technologies that establish authoritarian social relationships.

Toward DEMOCRATIC WORK:

B. Seek a diverse array of flexibly schedulable, self-actualizing technological practices. Avoid meaningless, debiliating, or otherwise autonomy-impairing technological practices.

Toward DEMOCRATIC POLITICS:

C. Seek technologies that can enable disadvantaged individuals and groups to participate fully in social and political life. Avoid technologies that support illegitimately hierarchical power relations between groups, organizations, or polities.

To secure DEMOCRATIC SELF-GOVERNANCE:

D. Keep the potentially adverse consequences (e.g., environmental or social harms) of technologies within the boundaries of local political jurisdictions.

E. Seek local economic self-reliance. Avoid technologies that promote dependency and loss of local autonomy.

F. Seek technologies (including an architecuture of public space) compatible with globally-aware, egalitarian political decentralization and federation.

To perpetuate DEMOCRATIC SOCIAL STRUCTURES

G. Avoid technologies that are ecologically destructive of human health, survival, and the perpetuation of democractic institutions.

exacerbated by the loss of public spaces (with, for instance, town commons being supplanted by shopping malls).[19]

Are there plausible alternatives? Zurich, Switzerland, has promoted a partial antidote by providing neighborhoods with legal advice and architectural assistance aimed at increasing community interaction.[20] Thanks to the program, neighbors have begun to remove backyard fences, to build new walkways, gardens, and other community facilities, and generally to refashion a system of purely private yards into a well-balanced blend of private, semi-public, and public spaces.[21]

A housing movement born in Denmark in the mid-1960s seeks, more ambitiously, to integrate desirable aspects of traditional village life with such contemporary realities as urbanization, smaller families, single-parent or working-parent households, and greater sexual equality. The result is "co-housing"—resident-planned communities ranging today from six to 80 households. More than 100 such communities now exist in Denmark and the Netherlands, and they are spreading to the United States and elsewhere.

The Trudeslund co-housing community, located near Copenhagen, comprises thirty-three families. Homes for each family cluster along two garden-lined pedestrian streets and are surrounded by ample open space and forested areas. Each home has its own kitchen, living room, and bedrooms, though these rooms have been somewhat downsized so that the savings can be used to construct and maintain common facilities. The latter include picnic tables, sand boxes, a parking lot, and, most importantly, a "common house" with a large kitchen and dining room, play rooms, a dark room, a workshop room, a laundry room, and a community store. Each night residents have the option of eating in the common dining room; cooking responsibilities rotate among all adults in the community (which means everyone cooks one evening a month). Because the community is designed to have residents walk past the common house on the way from the parking lot to any house, the common house becomes a natural gathering spot.

Trudeslund is successful by many measures. The common facilities save time and money, day care and baby sitting flow naturally from the pattern of community life, social interaction flourishes without sacrificing privacy, and safety and conviviality both prosper by banishing cars to the outskirts of the community. Over time cooperation has grown, with resident families choosing to purchase and share collectively tools, a car, a sailboat, and a vacation home. Rather than becoming insular, residents are actively involved in social and political life outside Trudeslund, with the common house serving as an organizing base for other activities.[22]

Insofar as mutual respect and equality are fundamental democratic values, an egalitarian community represents a democratic gain in its own right. Moreover, if one could envision creating an interacting network of such communities, one could expect to see greater respect, tolerance, and commonality emerging *between* communities, with beneficial implications for democratization on a broader scale.[23]

Criterion B: Democratic Work

Social scientists have hypothesized that the quality of our work life influences our moral development and our readiness to function as engaged citizens—that is, as active participants in a strong democracy.[24] Technology, in turn, plays a critical role in shaping our work experiences. Some years ago sociologist and one-time union organizer Robert Schrank discussed alternative work arrangements with a group of union representatives at the General Motors Corporation. After describing several experiments in Scandinavian factories that permitted more interesting work routines and greater worker involvement in the day-to-day decision-making, Schrank asked the men to imagine how they might redesign their own factories if given a chance. Their response was skeptical and unenthusiastic. Later Schrank reflected, "[T]he frame of reference of these workers was the linear assembly line as they experienced it. Even to think beyond that seemed difficult. . . ."[25]

Linear assembly lines not only tend to restrict possibilities for worker self-management, conviviality, and meaningful work but also to impair the ability of workers to envision technological alternatives. Schrank, however, was eventually able to show the GM workers more democratic automobile manufacturing technologies that have been in use for some years. For example, an innovative Volvo factory in Kalmar, Sweden, uses independently movable electronic dollies—each carrying an individual auto chassis—in place of a traditional assembly line. The dollies enable small teams of workers to plan and vary their daily routines for assembling automobile subsystems.[26]

A more creative and self-managed workplace is democratically desirable in itself. But it also can help workers develop the moral commitment, skills, and confidence to participate politically beyond the workplace.

Criterion C: Technology and Power

While political equality is essential for strong democracy, all contemporary political systems encompass groups whose opportunities for participating in social and political life are circumscribed. Today's technologies help reproduce this inequitable constellation of power. For instance, the technologies and architecture with which women must cope every day often help exclude them from the corridors of power. "Labor-saving" appliances "liberate" many wives to do housework that was once performed by other family members. (During the bygone era of open hearth cooking, for example, men chopped wood and children hauled water, thus contributing more equally to household maintenance.)[27] Likewise, modern neighborhood designs often isolate women socially, heighten their risk of physical abuse, and limit their opportunities to organize child care. The typical suburb lacks sidewalks, common gathering spaces, or the opportunity to work within a short distance of home.[28] Most public-transit systems have been designed without regard to women's typical social responsibilities. How is a mother supposed to get a baby carriage up onto a traditional bus or down the steps of a New York City subway station?[29] Many workplaces have jobs stereotyped as female that carry special risks of isolation, domination, stress, or harm. Secretaries and key-punch operators, who are preponderantly female, suffer unusually high levels of stress-related emotional and physical disorders. The marketing techniques of the mass media often degrade women and erect punishingly unattainable beauty standards.[30]

All of these consequences of technology limit women's opportunities to participate on equal terms in social and political life—including technological decisionmaking. To explain these results, one need not invoke theories of misogyny or conspiracy (although the temptation may be strong). Generally it seems more plausible to blame the indifference and insensitivity of male-dominated institutions and design professions, in which women's evaluations of their own needs rarely qualify as even a discussion topic.

Criterion D: Translocal Harms

Local self-governance is a key building block for strong democracy. The average citizen can exert much more influence locally than nationally, and local political equality and autonomy provide crucial opportunities for citizens to influence translocal politics.

Technologies can affect a community's ability to govern itself in several ways. For example, a technology that harms people in neighboring communities can provoke intercommunity conflict, which in turn can precipitate intervention by higher political authorities that subverts local self-governance. In the late 19th and early 20th century, American cities imported clean water, or filtered and treated incoming water, while discharging raw sewage into rivers and lakes. Various methods of sewage treatment were known or under development, but few cities adopted them (unless the raw sewage caused local harm). As the buildup of sewage increased illness and death in downstream communities, state governments passed preemptive laws protecting water quality, established state boards of health to help administer the laws, and created new regional governmental authorities ("special districts") charged with integrating and managing the systems of water supply and sewage treatment. The result of this state intervention was that water quality and public health dramatically improved—but local autonomy dramatically declined. Indeed, regional water management set a precedent that influenced the development of institutions governing transportation, electrification, and telephone communication. The failure of municipal governments to assume technological responsibility toward neighboring communities wound up subverting their own autonomy and the tradition of local self-governance.[31]

Today a related pattern continues to play out as large corporations repeatedly use cross-border pollution as a rationale to justify environmental regulation at ever higher levels of political aggregation (shifting, that is, from local to state, national, and ultimately international authorities). When "successful," this reallocation of power has transposed environmental decisionmaking to arenas relatively inaccessible to grassroots participation, where corporations have secured weak environmental standards that preempt stronger standards favored at the local level. This logic helps to explain industry support for the 1970 U.S. Clean Air Act and for the 1990 amendments to the Montreal Protocol (a treaty that regulates emissions of industrial chemicals hazardous to the earth's atmospheric ozone shield).[32] The erosion of local authority to control pollution has thus led to a violation of Criterion G in Table 1—"seek ecological sustainability."

Criterion E: Local Economic Self-Reliance

How can citizens meaningfully decide the fate of their community if economic survival depends on institutions or forces utterly beyond their control? Just as a measure of local political autonomy is essential for strong democracy, so, in turn, is a measure of local and regional economic self-reliance essential for political autonomy. Many modern technologies, however, can subvert self-reliance.

A century ago London differed from other leading world cities in eschewing reliance on a single major electric company, large generating plants, or even a city-wide electric grid. Instead there were dozens of small electric companies scattered throughout London—some privately owned, some public—deploying a diverse array of small-scale electrical generating technologies. Was London just backwards and irrational, as engineers from elsewhere commonly supposed? Not obviously. London's electric companies operated at a profit and provided reliable and affordable power adapted to local needs. London's borough governments, perceiving their own political significance and autonomy as inextricable from the infrastructures upon which they depended economically, consistently opposed Parliamentary efforts to consolidate the grid. The boroughs favored a highly decentralized electrical system that each could control more easily.[33]

For analogous reasons, a number of American cities, towns, and neighborhoods have begun to develop locally owned businesses oriented toward production for nearby markets. The Rocky Mountain Institute has developed an analytical process, along with supporting instructional materials, which a growing number of towns in economic difficulty are using to reduce their consumption of imported energy, water, and food (rather than to depend on distant supplies) and to reinvest local capital (rather than to put it in the hands of bankers thousands of miles away).[34] Once these communities are more secure against distant market forces or multinational corporate decisions, they are more empowered to conceive and undertake local democratic initiatives.[35]

This strategy of self-reliance contrasts strikingly with the prevalent strategy used by communities—self-defeating when it is not futile—of using concessionary tax breaks, waivers on environmental standards, or the promise of low wages to try to entice geographically fickle corporations. While generally decrying these corporate inducements, most proposed progressive technological strategies nonetheless remain preoccupied with advancing U.S. international competitiveness in ways that will assuredly *erode* local self-reliance.[36]

Democratic Design vs. Progressive Proposals

Several of the preceding examples suggest an important deficiency in the familiar progressive call to "rebuild America's crumbling technological infrastructure."[37] If our infrastructure needs repair or modernization, shouldn't we rebuild it in ways amenable to local democratic governance? For instance, there are technologies for managing industrial and municipal waste and conserving energy that can be deployed and administered with extensive local involvement.[38] Facilities to store solar energy for heating homes and buildings, as pioneered in Scandinavia, comprise another example of neighborhood-scale technology. Indeed, unless localities regain more control of their own infrastructures, there will be diminished incentive for the kind of grassroots political involvement essential to strong democracy. People only will participate in local politics when they have the power to affect important local decisions.

Another significant feature of all the preceding criteria is that they are designed to work together as a complementary system applied to an entire technological order. This too can improve the technology policies advocated by progressives. For instance, advocates of workplace democracy aim admirably to fulfill Criteria A and B, but generally fail to inquire whether the resulting goods and services are socially benign.[39] If we competitively and democratically produce democratically dubious technologies—say, chemical weapons, or certain consumer electronics like Walkmen that erode social interaction and solidarity—are we really making progress?[40]

Similarly, in an era of growing popular concern over acid rain, atmospheric ozone depletion, and global warming, few doubt the necessity of devising more ecologically sustainable technologies (Criterion G). Yet environmentalists sometimes imagine that sustainability alone is a sufficient basis for technological design.[41] To see the incompleteness here, recall the old sewage system configurations that both protected public health *and* subverted local self-governance, or consider Singapore's relatively stringent environmental policies that are coupled with a starkly authoritarian political regime.[42] In evaluating technology, we must learn to take into account all technologies, all their focal and non-focal effects, and all the manifold ways in which technologies influence political relations.

Toward a Democratic Politics of Technology

Democratic design criteria are essential to a democratic politics of technology, but only if coupled with institutions for greater popular involvement in all domains of technological decisionmaking. This suggests the need to establish new opportunities for popular participation to contest and apply the design criteria (in communities, workplaces, and other social realms), to set research and development (R&D) priorities, and to govern important technological systems. For instance, perhaps corporate R&D tax credits could be scaled up if a business introduces a democratic process or uses democratic design criteria to guide its R&D.

Our basic goal must be to open, democratize, and partly decentralize pertinent government agencies, to create avenues for worker and community involvement in corporate R&D and strategic planning, and to strengthen the capabilities of public institutions to monitor and, as needed, guide the political and social consequences of technology. We also need political strategies to accomplish these objectives, preferably built on popular movements and technological initiatives that already exist.[43]

One pertinent example of how to democratize technological decisionmaking began in the early 1970s, when natural gas was found beneath the frigid and remote northwest corner of Canada. Energy companies soon proposed building a high-pressure, chilled pipeline across thousands of miles of wilderness, the traditional home of the Inuit (Eskimos) and various Indian tribes. At that point a government ministry, anticipating significant environmental and social repercussions, initiated a public inquiry under the supervision of a respected Supreme Court justice, Thomas R. Berger.

The MacKenzie Valley Pipeline Inquiry opened its preliminary hearings to any Canadian who felt remotely affected by the proposal. Berger and his staff developed a novel format to encourage a thorough, open, and accessible inquiry. Formal, quasi-judicial hearings were held that combined conventional expert testimony with cross-examination. Berger also conducted a series of informal "community hearings." Traveling 17,000 miles to 35 remote villages and settlements, the MacKenzie Inquiry took testimony from nearly 1,000 native witnesses. And it provided funding to disadvantaged groups to support travel and legal counsel for more competent participation. The Canadian Broadcasting Company carried daily radio summaries of all the hearings in English and in six native languages.

One of the MacKenzie Inquiry's important lessons was that laypeople can produce useful social *and* technical information. According to one technical advisor:

> Input from nontechnical people played a key role in the Inquiry's deliberations over even the most highly technical and specialized scientific and engineering subjects ... [The final report] discusses the biological vulnerability of the Beaufort Sea based not only on the evidence of the highly trained biological experts who testified at the formal hearings but also on the views of the Inuit hunters who spoke at the community hearings. ... [Moreover,] when discussion turned to ... complex socioeconomic issues of social and cultural impact, [native] land claims, and local business involvement—it became apparent that the people who live their lives with the issues are in every sense the experts .... Their perceptions provided precisely the kind of information necessary to make an impact assessment.[44]

Quoting generously from expert and citizen witnesses, Berger's final report became a national best-seller. Within months the original pipeline proposal was rejected, and the Canadian Parliament instead approved an alternate route paralleling the existing Alaska Highway.[45] One can fault the MacKenzie Inquiry for depending so much on the democratic sensibilities and good faith of one man—Judge Berger—rather than empowering the affected native groups to play a role in formulating the conclusions. But the process was nevertheless vastly more open and egalitarian than comparable decisionmaking efforts in other industrial societies.[46]

There are many other prototypes for institutions or processes that enable greater popular involvement in technological decisions. For instance, the nascent Community Health Decisions (CHD) movement has developed grassroots procedures to forge popular consensus on ethical principles governing medical policy and technology. One of the accomplishments of the movement has been to organize dozens of community meetings throughout Oregon to debate proposed reforms in the state's health care system.[47] The CHD movement could provide a model for future forums in which citizens debate more general democratic design criteria for technology.

In several states, including Maine, Washington, and California, coalitions of peace activists, labor unions, business leaders, community groups, and government officials have created democratic processes to wean regional economies away from their dependence on military production. For example, responding to grassroots pressure, the state of Washington has established a citizen advisory group that monitors military spending in the state, assesses post-Cold War economic needs and opportunities, and promulgates action plans to help defense-dependent communities diversify their productive base.[48] This shows how a region can use social criteria to evaluate and redirect an entire technological order.

During the past twenty years the Dutch have developed a network of street corner "science shops," supported by nearby university staff and students, where citizen groups receive free assistance to address social issues with technical components. One science shop helped a local environmental group document the contamination of heavy-metals in vegetables, which pressured the Dutch government to sponsor a major clean-up in metalworking plants. The science shops have empowered citizens to participate in technological decisionmaking so successfully that they have prompted similar efforts throughout much of Western Europe.[49]

Traditional Amish communities, often misperceived as technologically naive or backwards, have pioneered popular deliberative processes for screening technologies based on their cumulative social impacts, in effect attending to many of the criteria listed in Table 1. One of their methods is to place the adoption of a new technology on probation for one year to discover what the social effects might be. For instance, Amish dairy communities in east central Illinois ran a one-year trial with diesel-powered bulk milk tanks before judging them socially acceptable; other Amish communities used social trials to prohibit once-probationary household telephones or personal computers.[50]

The preceding examples are, of course, atypical. Most technological choices are made by experts, bureaucratic machination, or unregulated market interactions. But the exceptions provide crucial evidence that, given the right institutional circumstances, lay citizens can make reasonable technological decisions reflecting their own priorities. Even federal agencies, such as the Office of Technology Assessment, the National Science Foundation, and the National Institutes of Health (NIH), have occasionally supported or incorporated citizen participation in their decisionmaking procedures.[51] For instance, the NIH has used both expert and lay advisory panels to evaluate research proposals. The experts judge the scientific merits, while laypeople help weigh the social value, political import, or ethical propriety. One can envision a wide range of private or public institutions using such models to develop a new system of democratic procedures for choosing among technological alternatives.

Participatory Research, Development, and Design

Democratic processes for technological choice and oversight are vital, yet hardly worth the effort unless participants have a broad range of alternative technologies from which to choose. Hence it is essential to weave democracy into the fabric of technological research, development, and design (RD&D).[52] Consider four examples of how this has been done.

Democratic Design of Workplace Technology

In Scandinavia during the early 1980s unionized newspaper-graphics workers—in collaboration with sympathetic university researchers, a Swedish government laboratory, and a state-owned publishing company—succeeded in inventing a form of computer software unique in its day. Instead of following trends toward routinized or mechanized newspaper layout, this software contained some of the capabilities later embodied in desktop publishing programs that enable printers and graphic artists to exercise considerable creativity in page design.[53] Known as UTOPIA, this project demonstrated how broadened participation in the RD&D process could lead to a design innovation that, in turn, supported one condition of democracy—creative work (Criterion B).

UTOPIA is less ambitious than several other attempts at participatory design within the workplace. For example, in the 1970s workers at Britain's Lucas Aerospace Corporation sought not only to democratize their own work processes but also to produce more socially useful products.[54] But UTOPIA demonstrated that workers could go beyond just developing prototypes. It also should be noted that this instance of collaboration between workers and technical experts—initially limited to a single technology within a single industry—occurred under unusually favorable social and political conditions. Sweden's workforce is 85 percent unionized, and the nation's pro-labor Social Democratic party has held power during most of the past 50 years.

Participatory Architecture

Compared with the relatively few examples of participatory design of machinery, appliances, or technical infrastructure, there is a rich history of citizen participation in architectural design. One example is the "Zone Sociale" at the Catholic University of Louvain Medical School in Brussels. In 1969 students insisted that new university housing mitigate the alienating architecture of an adjacent hospital. Architect Lucien Kroll established an open-ended, participatory design process that elicited intricate organic forms (e.g., support pillars shaped like gnarled tree trunks), richly diverse patterns of social interaction (e.g., a nursery school situated near administrative offices and a bar), and a dense network of pedestrian paths, gardens, and public spaces. Walls and floors of dwellings were movable, so that students could design their own living spaces. Construction workers were given design principles and constraints rather than finalized blueprints, and they were encouraged to create and display their own sculptures. Initially baffled by the level of spontaneity and playfulness, the project's structural engineers gradually adapted themselves to the diversity of competent participants.

Everything proceeded splendidly for some years until the university administration became alarmed at the extent to which they could not control the process. When the students were away on vacation, they fired Kroll and halted further construction.[55]

Feminist Design

What would happen if women played a greater part in RD&D? One answer comes from feminists who have long been critical of housing designs and urban layouts that reinforce the social isolation and the low, unpaid status of women as housewives.[56] If women were more actively engaged in community design, they might set up more shared neighborhood facilities for day care, laundry, or food preparation, or they might locate homes, workplaces, stores, and public facilities more closely together. Realized examples of feminist design exist in London, Stockholm, and Providence, Rhode Island.[57]

Another approach has been pioneered by an artist and former overworked-mother named Frances GABe, who devoted several decades to inventing a self-cleaning house. "In GABe's house," according to author Jan Zimmerman, "dishes are washed in the cupboard, clothes are cleaned in the closets, and the rest of the house sparkles after a humid misting and blow dry!"[58]

Other feminists have established women's computer networks and designed alternatives to dreary female office work and to transportation networks insensitive to women's needs.[59] An explicit feminist complaint against current reproductive technologies—such as infertility treatments, surrogate mothering, hysterectomy, and abortion—is that women have played a negligible role in guiding the medical RD&D agendas which have imposed on women agonizing moral dilemmas that might otherwise be averted or structured differently.[60]

Barrier-Free Design

During the past two decades there has been substantial innovation in the design of "barrier-free" equipment, buildings, and public spaces responsive to the needs of people with physical disabilities. Much of the impetus came from disabled citizens who organized themselves to assert their needs or helped invent design solutions. For example, prototypes of the Kurzweil Reading Machine, which uses computer voice-synthesis to read typed text aloud, were tested by over a hundred and fifty blind users. In an eighteen month period these users made over a hundred recommendations, many of which were incorporated into later versions of the device.[61]

Technology by the People

All these examples of participatory design demonstrate that it is possible to have a much wider range of people participating in technological research, development, and design.[62] Moreover, participatory design broadens the menu of technological choices. But many participatory design exercises also have encountered fierce opposition from powerful institutions—opposition engendered, not because the exercises were failing, but because they were succeeding.

Still, some advocates of participatory RD&D have elected to state their case entirely in terms of the material interests of the non-participants. Others have noted the contribution that participation can make to improved productivity or to better design solutions. These are all fair and reasonable arguments. What is rarely articulated is the specific moral argument that the opportunity to participate in RD&D should be a matter of right, because it is essential to individual moral autonomy, to human dignity, and to democratic self-governance.

A powerful moral case for participatory design has been made by people with disabilities who have demanded barrier-free design. The movement's achievements are now apparent in the profusion of ramps and modified rest rooms in public places (responsive to Criterion C), and they will soon become even more apparent with the promulgation of new regulations under the Americans with Disabilities Act of 1990. The movement not only opposes anti-democratic design but also has a constructive, hopeful thrust.

Non-market, democratic design criteria—often formulated and applied by disabled laypeople—are now being used to define individual and collective needs, including access to public spaces. Moreover, participants do not evaluate just one technology at a time—the norm in conventional technology assessment—but entire technological and architectural environments. When the range of democratic criteria broadens and when the participants expand beyond the disabled population, we will be well on our way toward ensuring that our technology is compatible with democracy.

Conclusion

Current technological orders are generally short on communitarian or cooperative activities, and long on isolation and authoritarianism (violating Criterion A). Work is frequently stultifying and tends to impair moral growth and political efficacy (violating Criterion B). Illegitimate power asymmetries are reproduced through technological means (violating Criterion C).

The opportunity to engage in a vibrant civic life is often preempted by shopping malls, suburban subdivisions, unconstrained automobilization, and an explosive proliferation in home entertainment devices. Thus we have diminished access to local mediating institutions or to public spaces that could support democratic empowerment within the broader society (violating Criteria A and F). The need to manage translocal harms, coupled with widespread dependence on centrally managed technological systems and with the growing integration of the global economy, has helped render local governments relatively powerless, thereby reducing anyone's incentive to participate (violating Criteria D, E, and F). Meanwhile, there is little compensating incentive to engage directly in national politics, which television reduces to a passive spectator sport, where powerful corporations exert disproportionate influence, where deep questions of social structure are slighted, and where the average citizen has negligible effect.

While it is not always easy to establish causal connections running from structural deficiencies to other social ills, it hardly seems conceivable that weak community ties, atrophied local political capabilities, and authoritarian and degraded work processes have had no influence upon illiteracy, stress, illness, divorce rates, teen pregnancy, crime, drug abuse, psychological disorders, and so on. Perhaps, as de Tocqueville foresaw, many of us *do* sometimes feel shut up in a narrow selfishness, robbed of one half of our existence, left with an immense void in our lives.

Progressive technological strategists face a dilemma. We can couch our nostrums in terms of prevailing economic goals like competitiveness and try to win short-run victories. Or we can strive for a world worthy of our ideals. But we can no longer pretend that the progressive policies so far proposed for improving national economic performance, any more than conservative policies, are going to avoid exacerbating the United States' most profound social and political maladies. Has not the time arrived to mobilize for a democratic politics of technology?

Notes

1. Alexis de Tocqueville, *Democracy in America,* ed. Phillips Bradley, rev. ed. (1848; reprint, New York: Vintage Books, 1945), Vol. 1, p. 260.
2. John J. Kushma, "Participation and the Democratic Agenda: Theory and Praxis," in Marc V. Levine *et al., The State and Democracy: Revitalizing America's Government* (New York: Routledge, 1988), pp. 14-48. According to the Harwood Group, many Americans would like to be more involved in public affairs, but feel locked out of the current system. See *Citizens and Politics: A View from Main Street America* (Dayton, OH: The Kettering Foundation, 1991).
3. Susan Friend Harding, *Remaking Ibieca: Rural Life in Aragon under Franco* (Chapel Hill: University of North Carolina Press, 1984).
4. Samuel Bowles and Herbert Gintis, *Democracy and Capitalism: Property, Community, and the Contradictions of Modern Social Thought* (New York: Basic Books, 1986). The converse causal tie between community strength and political empowerment is suggested, for instance, by solidaristic Amish communities' success in resisting mandatory public schooling, military conscription, and participation in the federal social security system. See Donald B. Kraybill, *The Riddle of Amish Culture* (Baltimore: Johns Hopkins University Press, 1989).
5. See, for example, Samuel Bowles, David M. Gordon, and Thomas E. Weisskopf, *Beyond the Wasteland: A Democratic Alternative to Economic Decline* (Garden City: Anchor Press/Doubleday, 1983); Stephen S. Cohen and John Zysman, *Manufacturing Matters: The Myth of the Post-Industrial Economy* (New York: Basic Books, 1987); Michael L. Dertouzos, Richard K. Lester, Robert M. Solow, and the MIT Commission on Industrial Competitiveness, *Made in America: Regaining the Productive Edge* (Cambridge, MA: MIT Press, 1989); Robert B. Reich, "The Real Economy," *Atlantic Monthly,* 267:2, February 1991, pp. 35-52.
6. See, for example, Joel S. Yudken and Michael Black, "Targeting National Needs: A New Direction for Science and Technology Policy," *World Policy Journal,* 7:2 (Spring 1990), pp. 282-3; Computer Professionals for Social Responsibility (CPSR), *The 21st Century Project: Funding Proposal* (Cambridge, MA: CPSR, 1991).
7. See also Langdon Winner, *The Whale and the Reactor: A Search for Limits in an Age of High Technology* (Chicago: University of Chicago Press, 1986).
8. For instance, public opinion polls reveal that American citizens' support for environmental protection has consistently far out-paced our political leaders' readiness to adopt stringent environmental policies. See Robert Cameron Mitchell, "Public Opinion and the Green Lobby: Poised for the 1990s?," in

Environmental Policy in the 1990s: Toward a New Agenda, eds. Norman J. Vig and Michael E. Kraft (Washington, DC: Congressional Quarterly Press, 1990), pp. 81-99. These findings suggest that under more democratic circumstances the United States would be much further along in addressing environmental issues.

9. See Albert Borgmann, *Technology and the Character of Contemporary Life: A Philosophical Inquiry* (Chicago: University of Chicago Press, 1984); Robert N. Bellah *et al., Habits of the Heart: Individualism and Commitment in American Life* (New York: Harper and Row, 1985); Alan Wolfe, *Whose Keeper? Social Science and Moral Obligation* (Berkeley: University of California Press, 1989); and Vandana Shiva, *Staying Alive: Women, Ecology and Development* (London: Zed Books, 1989).
10. See Ruth Schwartz Cowan, *More Work for Mother: The Ironies of Household Technology from the Open Hearth to the Microwave* (New York: Basic Books, 1983); and Borgmann, *Contemporary Life.*
11. Many Americans report dismay at the decline in face-to-face community. See Bellah *et al., Habits of the Heart;* The Harwood Group, *Citizens and Politics.* Yet few perceive technology's role in that decline, or the likelihood of concomitant adverse consequences for democracy. On the importance of community for democracy, see Benjamin Barber, *Strong Democracy: Participatory Politics for a New Age* (Berkeley: University of California Press, 1984).
12. During the mid-1980s annual U.S. sales of running shoes and of talking toy bears reached $2 billion and $100 million, respectively. The latter, smaller figure still "exceeded the value of all newly designed equipment promoted as appropriate technology sold in the Third World." Ken Darrow and Mike Saxenian, *Appropriate Technology Sourcebook: A Guide to Practical Books for Village and Small Community Technology* (Stanford, CA: Volunteers in Asia, 1986), p. 10.
13. I use the terms "focal" and "non-focal" to distinguish between any given technology's designed or intended purpose and that technology's accompanying complex of further, but often experientially recessive or less dramatic, functions, effects, attributes, and meanings. For example, a hammer's obvious focal function is to pound nails, but its many non-focal aspects include making noise, strengthening users' arms (and sometimes bruising their thumbs), and, in certain contexts, symbolizing justice.
14. See Sam Bass Warner, Jr., *Streetcar Suburbs: The Process of Growth in Boston (1870-1900),* 2nd ed. (Cambridge, MA: Harvard University Press, 1978); Samuel P. Hays, *American Political History as Social Analysis* (Knoxville: University of Tennessee Press, 1980); Christine Meisner Rosen, "Infrastructural Improvement in Nineteenth-Century Cities: A Conceptual Framework and Cases," *Journal of Urban History,* 12:3 (May 1986), pp. 211-56.
15. As examples, see Eric K. Drexler, *Engines of Creation: Challenges and Choices of the Last Technological Revolution* (Garden City, NY: Doubleday, 1986); U.S. Congress, Office of Technology Assessment, *Science, Technology, and the Constitution-Background Paper,* OTA-BP-CIT-43 (Washington, DC: U.S. Government Printing Office, 1987); or Jeffrey B. Abramson, F. Christopher Arterton, and Gary R. Orren, *The Electronic Commonwealth: The Impact of New Media Technologies on Democratic Politics* (New York: Basic Books, 1988).

16. Barber, *Strong Democracy*.
17. See, for example, Sara M. Evans and Harry C. Boyte, *Free Spaces: The Sources of Democratic Change in America* (New York: Harper and Row, 1986).
18. For the complete derivation and justification for these and additional democratic design criteria, see Richard E. Sclove, *Technology and Freedom: Toward a Democratic Politics of Technology, Architecture, and Design* (forthcoming).
19. See, for example, Kenneth T. Jackson, *Crabgrass Frontier: The Suburbanization of the United States* (New York: Oxford University Press, 1985); and Barber, *Strong Democracy*, pp. 267-73, 305-06.
20. I define technology broadly as material artifacts and the practices or beliefs that accompany their creation or use. Hence I regard architecture and community planning as a sub-domain of technology.
21. Dolores Hayden, *Redesigning the American Dream: The Future of Housing, Work, and Family Life* (New York: W.W. Norton, 1984), pp. 189-91.
22. Kathryn McCamant and Charles Durrett, *Cohousing: A Contemporary Approach to Housing Ourselves* (Berkeley, CA: Habitat Press, 1988).
23. Some supporting evidence can be found in the emerging international network of "sister cities." See, for example, Michael Shuman, "From Charity to Justice," *Bulletin of Municipal Foreign Policy*, 2:4 (Autumn 1988), pp. 50-59.
24. Edward S. Greenberg, *Workplace Democracy: The Political Effects of Participation* (Ithaca: Cornell University Press, 1986); William M. Lafferty, "Work as a Source of Political Learning Among Wage-Laborers and Lower-Level Employees," in *Political Learning in Adulthood: A Sourcebook of Theory and Research*, ed. Roberta S. Sigel (Chicago: University of Chicago Press, 1989), pp. 102-42; Melvin L. Kohn *et al.*, "Position in the Class Structure and Psychological Functioning in the United States, Japan, and Poland," *American Journal of Sociology*, 95:4 (January 1990), pp. 964-1008.
25. Robert Schrank, *Ten Thousand Working Days* (Cambridge, MA: MIT Press, 1978), p. 226.
26. *Ibid.*, 221-27. On remaining democratic shortcomings in the Volvo factories, see Stephen Hill, *Competition and Control at Work: The New Industrial Sociology* (Cambridge, MA: MIT Press, 1981), pp. 49 and 104-05. For further recent examples of both democratic and non-democratic workplace technology, see Shoshana Zuboff, *In the Age of the Smart Machine: The Future of Work and Power* (New York: Basic Books, 1988).
27. Cowan, *More Work for Mother*.
28. Hayden, *Redesigning the American Dream;* Ray Oldenburg, *The Great Good Place: Cafes, Coffee Shops, Community Centers, Beauty Parlors, General Stores, Bars, Hangouts, and How They Get You through the Day* (New York: Paragon House, 1989).
29. See Women and Transport Forum, "Women on the Move: How Public is Public Transport?" in *Technology and Women's Voices: Keeping in Touch*, ed. Cheris B. Kramarae (New York: Routledge & Kegan Paul, 1988), pp. 116-34.
30. Barbara Drygulski Wright, *Women, Work, and Technology: Transformations* (Ann Arbor: University of Michigan Press, 1987); Naomi Wolf, *The Beauty Myth: How Images of Beauty Are Used against Women* (New York: William Morrow, 1991).

31. See Joel A. Tarr, "Sewerage and the Development of the Networked City in the United States, 1850-1930," in *Technology and the Rise of the Networked City*, eds. Joel A. Tarr and Gabriel Dupuy (Philadelphia: Temple University Press, 1988), pp. 159-85; and Gerald E. Frug, "The City as a Legal Concept," *Harvard Law Review*, 93:6 (April 1980), pp. 1057-1154.
32. See Samuel P. Hays, *Beauty, Health, and Permanence: Environmental Politics in the United States, 1955-1985* (Cambridge, England: Cambridge University Press, 1987), pp. 443-5, 456-7; Gareth Porter and Janet Welsh Brown, *Global Environmental Politics* (Boulder, CO: Westview Press, 1991), pp. 64, 66; Wolfgang Sachs, "Environment and Development: The Story of a Dangerous Liaison," *The Ecologist*, 21:6 (November/December 1991), pp. 252-57. Local ability to pressure corporations to reduce pollution could be much advanced by supportive legislation such as the Environmental Bill of Rights proposed by Bowles *et al.*, *Beyond the Wasteland*, pp. 344-6.
33. Thomas Parke Hughes, *Networks of Power: Electrification in Western Society, 1880-1930* (Baltimore: Johns Hopkins University Press, 1983), Chapter 9.
34. See Robert Gilman, "Four Steps to Self-Reliance: The Story Behind Rocky Mountain Institute's Economic Renewal Project," *In Context*, 14 (Autumn 1986), pp. 41-6; Barbara A. Cole, *Business Opportunities Casebook* (Snowmass, CO: Rocky Mountain Institute, 1988); and David Morris, "Self-Reliant Cities: The Rise of the New City-States," in *Resettling America: Energy, Ecology and Community*, ed. Gary J. Coates (Andover, MA: Brick House Publishing Co., 1981), pp. 240-62.
35. See John Gaventa, *Power and Powerlessness: Quiescence and Rebellion in an Appalachian Valley* (Urbana: University of Illinois Press, 1980), Chapter 8; and Frug, "The City as a Legal Concept."
36. See, for example, Cohen and Zysman, *Manufacturing Matters; Dertouzos et al., Made in America;* Lester C. Thurow, *The Zero-Sum Solution: Building a World-Class American Economy* (New York: Simon & Schuster, Inc 1985); and Yudken and Black, "Targeting National Needs." While sharply critical of economic nationalism, Robert B. Reich's *The Work of Nations* (New York: Vintage Books, 1992) assumes increased integration into an ever more intensively globalized economy. On the importance of granting greater *local* power over national self-reliance, see Ann J. Tickner, *Self-Reliance versus Power Politics: The American and Indian Experiences in Building Nation States* (New York: Columbia University Press, 1987).
37. See, for example, Bowles *et al., Beyond the Wasteland;* Yudken and Black, "Targeting National Needs;" and Reich, "The Real Economy."
38. Amory B. Lovins, *Soft Energy Paths: Toward a Durable Peace* (Cambridge, MA: Ballinger, 1977); National Center for Appropriate Technology, *Wastes to Resources: Appropriate Technologies for Sewage Treatment and Conversion*, DOE/CE/15095-2 (Washington, DC: U.S. Government Printing Office, July 1983); Darrow and Saxenian, *Appropriate Technology Sourcebook;* Valjean McLenighan, *Sustainable Manufacturing: Saving Jobs, Saving the Environment* (Chicago: Center for Neighborhood Technology, 1990).

39. See, for example, Michael J. Piore and Charles F. Sabel, *The Second Industrial Divide: Possibilities for Prosperity* (New York: Basic Books, 1984); Cohen and Zysman, *Manufacturing Matters.*
40. David F. Noble's deservedly influential essay, "Social Choice in Machine Design: The Case of Automatically Controlled Machine Tools," lauds Norwegian factory worker involvement in technology choices that helped workers maintain autonomy and creativity. True enough, but the factory in question was a state-owned weapons production plant. In *Case Studies on the Labor Process,* ed. Andrew Zimbalist (New York: Monthly Review Press, 1979), pp. 18-50.
41. See, for example, John Todd and Nancy Jack Todd, *Bioshelters, Ocean Arks, City Farming: Ecology as the Basis of Design* (San Francisco: Sierra Club Books, 1984).
42. Stan Sesser, "A Reporter at Large: A Nation of Contradictions," *The New Yorker,* 13 January 1992, pp. 37-68.
43. See Richard E. Sclove, "The Nuts and Bolts of Democracy: Toward a Democratic Politics of Technological Design," in *Critical Perspectives on Non-Academic Science and Engineering,* ed. Paul T. Durbin (Bethlehem, PA: Lehigh University Press, 1991), pp. 239-62; and Sclove, *Technology and Freedom.*
44. D.J. Gamble, "The Berger Inquiry: An Impact Assessment Process," *Science,* 199:4332 (3 March 1978), pp. 950-51.
45. *Ibid.,* pp. 946-52; Thomas R. Berger, *Northern Frontier, Northern Homeland: The Report of the MacKenzie Valley Pipeline Inquiry,* 2 vols. (Ottawa: Minister of Supply and Services, Canada, 1977); Organisation for Economic Cooperation and Development (OECD), *Technology on Trial: Public Participation in Decision-Making Related to Science and Technology* (Paris: OECD, 1979).
46. See, for example, Barry M. Casper and Paul David Wellstone, *Powerline: The First Battle of America's Energy War* (Amherst: University of Massachusetts Press, 1981); David Dickson, *The New Politics of Science* (Chicago: University of Chicago Press, 1988).
47. Bruce Jennings *et al.,* "Grassroots Bioethics Revisited: Health Care Priorities and Community Values," *Hastings Center Report,* 20:5 (September/October 1990), pp. 16-23.
48. Kevin J. Cassidy, "Defense Conversion: Economic Planning and Democratic Participation," *Science, Technology, and Human Values,* 17:3 (Summer 1992), pp. 334-48.
49. Seth Shulman, "Mr. Wizard's Wetenschapswinkel," *Technology Review,* 91:5 (July 1988), pp. 8-9.
50. On Amish technological decisionmaking see, for example, Marc A. Olshan, "Modernity, the Folk Society, and the Old Order Amish: An Alternative Interpretation," *Rural Sociology,* 46:2 (Summer 1981), pp. 297-309; Victor Stoltzfus, "Amish Agriculture: Adaptive Strategies for Economic Survival of Community Life," *Rural Sociology,* 38:2 (Fall 1973), pp. 196-206; Kraybill, *The Riddle of Amish Culture.*
51. See, for example, U.S. Congress, Office of Technology Assessment, *Coastal Effects of Offshore Energy Systems* (Washington, DC: U.S. Government Printing Office, 1976); Rachelle Hollander, "Institutionalizing Public Service Science: Its Perils and Promise," in *Citizen Participation in Science Policy,* ed. James C. Petersen (Amherst: University of Massachusetts Press, 1984), pp. 75-95.

52. For additional arguments in support of participatory design, see Richard E. Sclove, "The Nuts and Bolts of Democracy: Democratic Theory and Technological Design," in *Democracy in a Technological Society*, ed. Langdon Winner (Dordrecht: Kluwer Academic Publisher, 1992), pp. 139-157.
53. Andrew Martin, "Unions, the Quality of Work, and Technological Change in Sweden," in *Worker Participation and the Politics of Reform*, ed. Carmen Sirianni (Philadelphia: Temple University Press, 1987), pp. 95-139.
54. Hilary Wainwright and Dave Elliott, *The Lucas Plan: A New Trade Unionism in the Making?* (London: Allison and Busby, 1982).
55. Lucien Kroll, "Anarchitecture," in *The Scope of Social Architecture*, ed. Richard C. Hatch (New York: Van Nostrand Reinhold, 1984), pp. 166-85.
56. Hayden, *Redesigning the American Dream*, Chapter 4.
57. *Ibid.*, pp. 163-70.
58. Jan Zimmerman, *Once Upon the Future: A Woman's Guide to Tomorrow's Technology* (New York: Pandora, 1986), pp. 36-7.
59. Wright, *Women, Work, and Technology*; Kramarae, *Technology and Women's Voices*.
60. See Sarah Franklin and Maureen McNeil, "Reproductive Futures: Recent Literature and Current Feminist Debates on Reproductive Technologies," *Feminist Studies*, 14:3 (Fall 1988), pp. 545-60.
61. See Michael Hingson, "The Consumer Testing Project for the Kurzweil Reading Machine for the Blind," pp. 89-90, and Raymond Kurzweil, "The Development of the Kurzweil Reading Machine," pp. 94-96, in Virginia W. Stern and Martha Ross Redden, eds., *Technology for Independent Living: Proceedings of the 1980 Workshops on Science and Technology for the Handicapped* (Washington, DC: American Association for the Advancement of Science, 1982).
62. For a number of additional examples, see Sclove, "The Nuts and Bolts of Democracy: Toward a Democratic Politics of Technological Design.."

Chapter 3

Technology and Control in the Workplace

by Charley Richardson

For the vast majority of Americans, the workplace is one of the most important institutions in daily life. It is where we spend a great portion of our time as adults and earn our only major source of income. The workplace and work affect our circle of friends, our social status, our lifestyle, our health, and our children's future. Today, new technologies and new forms of work organization are unleashing major changes in America's workplaces. Technology obviously has an impact in other social arenas such as the home and the military, but its impact is especially powerful in the workplace, where there are minimal mechanisms for input and control by those who are most directly affected—workers.

This chapter explores how technology is being used in the workplace, where current technological trends are heading, and what is needed to redirect technology toward a more equitable society. Ultimately, these issues raise a profound question of democracy: Should the majority have a voice in decisions that directly affect their lives? Because of the imbalance of power in technological arenas and the declining level of unionization in the United States, there is a need for radical reform if the voice of the workforce is to be heard.

The arguments and examples put forth in this paper come primarily, though not exclusively, from the experiences of workers in the manufacturing sector. However, it should be noted that the issues raised here are relevant to workers in all sectors of the economy. Computerized learning, for example, threatens the role of teachers as skilled professionals as much as computer-controlled tools threaten the role of machinists.

For many people, the word technology brings to mind images of computers, robots, and hardware. But to understand the impact of workplace technology, a broader definition of technology is necessary. Technology is more appropriately understood as the embodiment of knowledge, attitudes, and culture in productive activity. It includes not only hardware but also software, materials, techniques, and work organization. Technology is what we use to create products and to provide services.

The traditional view of technology as hardware is neither insignificant nor coincidental. That definition serves a social purpose by limiting people's sense of control over the development and application of technology. Machines are fixed structures and seemingly unchangeable. But in fact, technology of any sort is the end result of a series of choices made by people. Technology both reflects social relations and reinforces them. Consider, for example, something as simple as a hand tool. In many industrial jobs, tools are designed for the average male hand and are too large for the average female hand. The tool thus reflects a social order which excludes women. And by making success on the job more difficult for women, the tool also reinforces a male-biased social order.

The organization of work is a particularly important and often ignored type of technology. Indeed, the social impact of many technologies occurs primarily through their effect on the workplace. A piece of equipment or machinery (even with a social bias built into it) has little significance until it is used in production. In the hand-tool example, work could be organized to counteract the exclusionary design or it could be organized to enhance it. Once a piece of hardware or equipment becomes part of the production process, it becomes part of a social structure and ceases to be a value-free "thing." A Computer Numerical Control (CNC) machine tool sitting in a parking lot and doing nothing has no social impact. The same tool being used to increase productivity can eliminate jobs, transfer skills off the shop floor, and set off economic and social tremors.

The impact of a particular piece of hardware depends on how it is applied. Indeed, the social impact of many other technologies occurs through their effect on the organization of work. For example, traditional machinists had considerable skills and made numerous decisions on the shop floor. Because of their relatively high degree of autonomy, they had relatively high wages. But if installation of a CNC machine moves decision-making into a programing department, the machinists' skills and autonomy diminish, their wages drop, and their jobs become less rewarding. The CNC also diminishes the ability of individual machinists and their union to negotiate over working conditions, such as break time, production quotas, and health and safety standards. But the same piece of equipment also could be used in a way that leaves the programming function in the hands of the machinists, allows machinists to acquire new skills, and enhances their social position and economic power.

The Origins of Workplace Technologies

None of us would like to hold many of the jobs created by workplace technologies. Assembly line work, for example, involves highly repetitive activity, putting the same part onto the same product, over and over again, day in and day out. Cycle times (times between the repetition of tasks) of one minute or less are not uncommon on contemporary assembly lines. The opportunities to take needed breaks, to apply intelligence, or to be more than a cog in a machine are very limited.

While the majority of Americans do not work on assembly lines, the same principles of technology development and workplace design that were popularized in Charlie Chaplin's movie "Modern Times" are found in every sector of the economy. For example, the letter sorting machine (LSM) in the U.S. Post Office requires repetitive, assembly-line-type tasks. Operators of the LSM's sit at a console and sort letters as they fly by at the rate of one-per-second. Postal workers must read the address and type in a three or four digit code, which is used by the machine to direct the letter to the appropriate bin. The operator has no ability to adjust the pace of the machine.

A directory assistance operator, using modern technology, must perform equally repetitive tasks. Operators are monitored continuously and required to keep their calls within a certain average amount of time. With today's technology, the task of telling the number to the customer, once the easiest and most satisfying part of the job, has been taken away and given to a voice synthesizer. As a result of this automation, the operators are given more calls to handle rather than additional breaks.

Why are jobs becoming increasingly inhuman? Certainly not because workers want them to be so. Jobs have been designed in this country according to a set of precepts called "scientific management," or Taylorism. Frederick Taylor developed this system in the late 1800s with the specific goal of removing control over the work process from the craft workers and placing it firmly in the hands of management.

Taylor's name and system are often connected with efforts to improve efficiency and productivity. But even his official (and sympathetic) biographer harbored no confusion about Taylor's real goal: "Back of all this intellection [sic] of his can be seen the grand aim of *control.* The importance he placed on control cannot be exagerated [sic]."[1] Taylor's method for gaining control over the production process was to remove all decision-making from the workers, to separate planning from execution, and to downgrade work by breaking it down into its smallest possible components, a process he called microdivision. Knowledge was to be extracted and controlled by management, and the actions of the workforce were to be strictly regulated. As Harry Braverman describes, "...it is thenceforth [the duty of workers] to follow unthinkingly and without comprehension of the underlying technical reasoning or data."[2] The downgrading of work was based on a de-skilling of the workforce. But Taylor went further and promoted a downgrading of the workforce itself. In his early book, *Shop Management*, Taylor frankly stated that the "full possibilities . . . will not have been realized until almost all of the machines are run by men who are of smaller calibre and attainments, and who are therefore cheaper than those required under the old system."[3] Taylor and his followers sought not only people without skills but also people of "lower potential," who would fit more naturally into degraded jobs and who would not expect higher pay.

The economic effectiveness of Taylor's approach depended on three assumptions: that only management can determine the "one best way" to accomplish a given task; that shop-floor knowledge and learning are irrelevant; and that to a certain extent, innovation and change are not needed. Taylor and his disciples attempted to eliminate any reliance on workers' skills, ideas, and creativity. They saw management control, not worker input, as the path to progress. According to his biographer, Taylor "provided for progress when, in defining the functions of management, he said that 'one man should be especially charged with the work of improvement in the system and in the running of the plant."[4]

Taylor attempted to sell his system to both labor and management by suggesting that there were objective means to determine the appropriate means of production, the amount of labor needed, and the right pay levels. Scientific analysis, he argued, could eliminate all labor-management confrontation and generate productivity gains that would make a better life for all.

> The amount of work which a man should do in a day, what constitutes proper pay for this work, and the maximum number of hours per day which a man should work . . . can be much better determined by the expert time student than by either the union or a board of directors, and . . . in the future scientific time study will establish standards which will be accepted as fair by both sides.[5]

Current engineering texts extol Taylor's methods. "With Taylor, as with managers today," explains one such tome, "time study was a tool to be used in increasing the over-all efficiency of the plant, making possible higher wages for labor and lower prices of the finished products to the consumer."[6] This argument is hauntingly similar to the current rhetoric about "new approaches" to production that promote statistical (objective, mathematical) control and "win-win" gains for both labor and management.

Despite the suggestions of broad social benefits, Taylor's work was driven by a conscious and openly stated desire to overcome the workforce's tendency to watch out for its own interest. He attempted to create a production system that used only the physical and readily measurable attributes of work, that limited workers' freedom of choice, that devalued their knowledge, and that actively sought to downgrade both jobs and the workers who performed them. He understood well the contradiction between worker skill and management control. The perfect worker, from Taylor's point of view, was simple-minded, physically adept, and eminently unquestioning.

Although Taylorism and scientific management are often considered in the context of narrow, repetitive work, the values they engendered have been used to measure and control skilled work as well. While I was working as a skilled heavy steel fabricator (a shipfitter) at the Sun Shipbuilding and Drydock Company in the late 1970s, a time study engineer was assigned to analyze my job. He performed random observations of my work over an eight-hour period and drew up a profile of my working day, with percentages for different activities, including taking a break. I later asked the engineer, if he could tell whether, while I stood still beside the unit I was working on, I was actively thinking about the job or merely day-dreaming. His answer, of course, was that he could not tell. He assumed I was "doing nothing." He assumed that thinking was not part of my job. My most important contribution to the production process was rejected as non-existent. The system of measurement, and the reward system connected to it, rejected innovation, thought, and planning by skilled workers.

In the end Taylorism has served and continues to serve the perceived interests of management against the needs and interests of the workforce. It has been and remains a system designed to give management control over the method, quantity, and pace of production. It has quite deliberately designed production systems—that is, technologies—based on repetitive tasks with limited mental content.

Taylorist work design, and the problems it engenders for workers, is widespread. In Sweden, one of the countries that has made the greatest strides in moving away from Taylorism, the metalworkers union found that 40 percent of its membership is still engaged in work that has a cycle time of less than five minutes.

Taylorism is not benign. The highly repetitive nature of jobs has created an epidemic of repetitive strain injuries (RSI) such as Carpal Tunnel Syndrome. Worker stress is also a growing problem, a direct result of jobs that place high demands on workers while giving them very little control. Stress from Taylorized jobs has been linked to an increased incidence of coronary heart disease and to other significant health problems.[7]

New Technology

How is Taylorism manifested by production technologies today? Despite widespread discussion of employee empowerment, worker involvement, and high-skill production strategies, management still seems eager to use technologies that minimize reliance on the skills of the workers. Even those who speak glowingly and nostalgically about skill enhancement often seek to diminish it.

For example, in a book called *Manufacturing Intelligence,* researchers from Carnegie Mellon University argue that in small batch manufacturing, which is the basis of a high value-added competitive industry, "the availability of traditional craftsmanship is of utmost importance because many unforeseen situations are encountered."[8] They point out that "several factors have led to a severe shortage of skilled machinists in vital industries . . . ." Historically, they say, problems are created and opportunities missed because managers tend "to undervalue the importance of the skilled craftsman's role on the shop floor." These problems are significant:

> We appear to be in a dangerous position in manufacturing history. A variety of societal factors have combined to make these skilled machinist craftsmen in short supply; yet at the same time, consumer pressures and national defense issues are driving manufacturing to small batch sizes. Here, the role of craftsmanship is vitally important.[9]

The authors seem to argue that skill and skilled workers are important and that Taylorist job design is ultimately bad for the economy. One would certainly think that better training programs and more respect for and better use of worker skill might help. But the authors have another solution: "We conclude, therefore, that craftsmanship must be automated and that *machine tools must be made more intelligent.*" Despite their pronouncements about the importance of skill in today's changing world, they, like Taylor, seek a means to separate skill from the workforce.

While leading-edge universities try to find ways around the use of skilled workers, even as they extol the virtues of craft, machine-tool manufacturers try to sell systems that separate planning from execution and that rely on low-skill (and inexpensive) labor. Consider a recent advertisement by a major machine tool manufacturer: "The TECHMASTER is a three-axis CNC system for surface, profile, and creep-feed grinding. It gives the manufacturing engineer complete control over the grinding process, *eliminating the need for highly skilled operators.*"[10] These trends are not only occurring in the manufacturing setting. A major producer of medical record tracking systems for hospitals claimed at a sales demonstration, for example, that its software is so simple that a "mentally retarded person could use it."

An inevitable outcome of these attempts to avoid reliance on skilled workers is that training systems are allowed to deteriorate.[11] It is no accident that the United States has one of the best university systems in the world and one of the worst training systems for its working people.[12] This disparity reflects the Taylorist view that progress comes only from scientific knowledge and that workers are to be controlled rather than empowered. It reflects the weakness of existing mechanisms for worker input.

While improving the skills of the workforce is currently a hot topic, actual training programs, to the extent that there are any, suffer from a narrow view of workers' needs and capabilities. The fundamental problem is that these programs seek to meet management's needs. Businesses are caught in the contradiction between their demand for creativity and their need for control. Much of what masquerades for training is either directed toward very basic skills like reading and writing or toward cultural or attitudinal training (that is, winning workers over to management's point of view).

The Congressional Office of Technology Assessment explained in a recent report, for example, how Statistical Process Control (SPC) training is used to move the workforce in a particular direction. Companies train shopfloor workers in SPC methods in large part to impress upon them the need for continuous and disciplined attention to their work. Rarely do they expect employees to learn anything about statistics beyond a few simple terms like averages and ranges.[13] Rather than reaffirming the importance of a skilled workforce, this narrow approach to training downgrades skill in a way that would thrill the practitioners of Taylorism.

The following passage is from a publication called *Workplace Basics: The Skills Employers Want*:

> Once the status leader and central figure on the working team, the machinist applied various technologies to shape individual parts of a final product. Career advancement once depended on sharpening the essential job skill of hand-to-eye coordination. Eventually, when the machinist's hand-to-eye coordination skills were good enough, he or she became a tool and die maker. Enter the robot with more consistent hand-to-eye coordination than the machinist, a better performer of the basic job task, [and] the robot took the machinist's place.[14]

Again the contradiction surfaces between reliance on and fear of skill. No one who respected machining or machinists would argue that the essential job skill of the machinist is hand-eye coordination. This document, written in 1988, was funded by the U.S. Department of Labor. Its conclusions laid the basis for development of national and local training policies and programs. Overcoming the current disrespect for skilled workers is a prerequisite for decent training programs.

Japanese work systems are often praised as embodying a different, anti-Taylorist approach to technology. Indeed, some Japanese industrialists have identified the failure of innovation that results from Taylorism as one of the key explanations for the decline of U.S. manufacturing. Konosuke Matsushita, founder of Matsushita Electric Industrial Co., put it succinctly:

> Survival is very uncertain in an environment increasingly filled with risk, the unexpected, and competition. Therefore, a company must have the constant commitment of the minds of all its employees to survive . . . .We have measured—better than you—the new technological challenges. We know that the intelligence of a few technocrats—even very bright ones—has become totally inadequate to face these challenges. Only the intellects of all employees can permit a company to live with the ups and downs and the requirements of its new environment. Yes, we will win and you will lose. For you are not able to rid your minds of the obsolete Taylorisms that we never had.[15]

This Japanese manufacturing leader certainly understands, particularly in the context of today's changing markets and changing technologies, that Taylor's rejection of worker knowledge was wrong-headed. But he also is sympathetic with Taylorism's fundamentally anti-democratic message—that is, its promotion of management control. There is no confusion about whose interests should be served by the knowledge of the workforce. Matsushita goes on to speak of the need to mobilize "the entire workforce's intellectual commitment *at the service of the company.*" Again, the workers' interests are made subservient to those of the company. Workers are expected to use their creativity and knowledge, but not for their own interest. This persuasion is generally achieved through cultural mechanisms that convince workers that their interest and the company's are one and the same. The ironic part is that the workers' ideas then are used in ways that eventually Taylorize work even further.[16]

One might ask why, if current technologies are so destructive of those who work with them, are not other types of technology developed? One answer is that Taylorism is deeply imbedded in our engineering systems. To explore this issue more thoroughly, several colleagues and I developed the following discussion case-study for engineering courses:

> Uncle Bell is about to install a new, computerized system for its telephone operators. Recent internal studies have shown the need to increase efficiency by eliminating supervisors and cutting wasted time for operators. Today, a supervisor must collect and interpret computer data to determine if individual operators are meeting the required performance level. This process, however, is time consuming and merely reports poor efficiency. It does not improve efficiency.
>
> Our engineers have developed a system which will solve these problems. We will be placing in every operator headset a tiny transmitter that is capable of producing a mild electric shock. When the computer objectively determines that an individual is not maintaining her/his required rate, a shock will be administered. Thus, inefficient operators will be instantly notified. Our initial tests have proven that this shock poses no threat of physical harm to the operator. Through the use of this system we will be able to provide better and less expensive service to our customers.

The response to this case is enlightening. While all the engineers in the classes we taught rejected this method of increasing efficiency, they were sharply divided in their reasoning. About half did so on simple moral grounds. The other half were clearly uncomfortable with the system, but felt a need to cover their values with an "objective" analysis of the inefficiency of the system. More interesting was the fact that none of the engineers felt the same moral outrage about the design of jobs under Taylorism that are just as destructive of human potential, that create stress-related health problems, and that lead to disabling repetitive-strain injuries.

The pathway for the development of technology is apparently seen as fixed, as waiting to be discovered. The possibility that there exists a wide range of technological options is essentially unthinkable to those raised under the current system. This is a part of the mystification of technology that can be seen in three other widely held views: that technology is objective and not subject to social control; that technological progress is inevitable and cannot be affected; and that engineering is a field exclusively for experts in which common people cannot participate. This mystification, combined with the ideology of private property rights, strips workers of any access to technological decisionmaking. Instead, decisionmaking is relegated to a small, homogenous group that is primarily white and male.

New Technology, New Possibilities

Can things be different? And if so, what would "different" look like? For many Americans, including some who consider themselves politically "progressive," the key questions seem to be about how to achieve competitiveness, efficiency, and profitability. I would suggest that a better focus for a technology discussion would be on how to create good jobs. This discussion would have to start with the question: what is a good job?"

When my colleagues and I present this question to people, respondents mention not only a decent income but also respect, variety, challenge, security, health, safety, social interaction, and control over the work process. Unfortunately, there are essentially no mechanisms for these preferences to be communicated to (or legally imposed upon) those who develop technologies for the workplace. Instead, the disciples of Taylor build technologies around a set of values and goals that are antithetical to what workers deem a good job. Engineers and others have learned to live with the contradiction between what they want for themselves and what they create for others.

One of the ways in which the designers of workplace technologies come to terms with these contradictions, a favorite throughout history, is to define the victims as members of a different and lower caste. For example, in an industrial hygiene class for engineers, a student's suggestion that jobs be analyzed according to his classmates' standards for desirable employment is met with derision, on the assumption that workers do not want, need, or deserve the same things as do engineers and other professionals.

This is not to say that Taylorists seek to damage workers. Taylorists no more wish to damage employees than the people who knowingly designed and produced the Ford Pinto intended to burn to death its passengers. But Taylorists nevertheless ignore the clear evidence that their design principles are hurting people.

Good jobs must be the starting point for the development of desirable workplace technologies. Once society provides its citizens with good jobs, other goals such as a strong sense of community, decent health care, and environmental cleanup become possible. Good jobs mean a strong tax base. They mean employees with fewer health and psychological problems. They mean a lower incidence of poverty, crime, despair, and suicide.

In 1964, the original Triple Revolution statement recognized that work is an important mechanism for income distribution and called for full-employment measures to counteract the job-destroying effects of cybernation and technological change:

> The fundamental problem posed by the cybernation revolution in the United States is that it invalidates the general mechanism so far employed to undergird people's rights as consumers. Up to this time economic resources have been distributed on the basis of contributions to production, with machines and men competing for employment on somewhat equal terms.[17]

"The Triple Revolution" also argued that the "underlying cause of excessive unemployment is the fact that the capability of machines is rising more rapidly than the capacity of many human beings to keep pace."[18] But the statement did not focus on the social role of work—on the importance of the *kind* of work people do—to their lives, to their self-esteem, to their interaction with society.

"The Triple Revolution" envisioned an important and expanded role for trade unions by recommending all of the following:

> a. Use of collective bargaining to negotiate not only for people at work but also for those thrown out of work by technological change.
>
> b. Bargaining for perquisites such as housing, recreational facilities and similar programs . . . .
>
> c. Obtaining a voice in the investment of the unions' huge pension and welfare funds . . . .
>
> d. Organization of the unemployed . . . and strengthening of the campaigns to organize white-collar and professional workers.[19]

These are still valid goals for trade unions, but other goals are equally essential. The *quality* of jobs matters as much as the number. Most jobs in our society steal from people rather than enrich their development. Some jobs develop the knowledge and creativity of workers, while others stifle and deaden these capacities. Some jobs promote a healthy lifestyle, while others expose people to repetitive-motion injuries, stress-related disorders, and dangerous chemicals. The issue is not just how to create jobs and income but also how to create good jobs.

While "The Triple Revolution" envisioned too narrow a role for trade unions, the oversight reflected a more general problem—that trade unions have been denied a role to represent workers' interests within the process of technology development and implementation. Unions have not been allowed to negotiate over technology itself.

Input vs. Control

The rapidly changing nature of technology has undermined the ability of trade unions to protect the interests of their members. Standard three-year contracts, for example, can actually undermine the position of workers when technology is changing every year or every month. Contract language on issues of technology is generally limited to how to dispose of workers made redundant by technological change.

There is currently a great deal of talk in the popular and business press about worker participation in decisions concerning technology. But does this represent a real redistribution of power and a move toward democratic control of the workplace? While businesses today recognize that the ideas of workers are very important, a closer look at most worker participation programs reveals that control remains in the hands of management.

Many decisions about which technologies will be used for production are made during research-and-development stages that are entirely immune to worker input. Moreover, three other tough questions must be asked to understand the power dynamics within participatory programs: Who has access to resources like information? Who sets the standards for behavior and for evaluation? Who defines the problem?

Asking who has access to information reveals who can exert power within a decisionmaking process. Very often, when unions become engaged in discussions over technology, they have neither the information nor the training that would allow them to play an equal role. Management generally controls all the information about future plans and options.

Management also controls the standards of behavior and evaluation. This, by itself, limits the value of worker input. If efficiency and productivity, defined from the company's perspective, are accepted as the standards by which new technology is judged, then the involvement of workers will do little to change the types of technology that are chosen. When workers at a shipyard under a QWL (quality of worklife) program recently designed new tools that could make their jobs easier, they were only given clearance to use the tools after they had proved to management that the innovations would save the company money. If the tools had only made their jobs more satisfying or had not provided financial gain for the company, permission to purchase them might have been denied. The workers' interest in better work does not have even equal footing with management's interest in cutting costs.

Another example of the importance of setting standards comes from Japan, where many companies require workers to wear uniforms. These dress codes are justified and promoted as standards of equality, but in fact they are standards of conformity in the view of management. Union buttons and other statements of labor solidarity are banned.

A third example of a company-imposed standard is a negative view of absenteeism. Workers in participatory situations are often pushed to socially sanction their co-workers when they miss work. By another, more humane set of standards, staying home to fully recover from an illness or to care for a sick child might be promoted rather than penalized.

Finally, there's the question of who defines the problem. Today we hear a lot about the need for problem-solving by workers but very little about democratic mechanisms to decide which problems need to be solved. Lack of competitiveness is often described as the core problem in our economy, rather than the lack of good jobs and the inequitable distribution of resources. Most worker-participation programs allow discussions about how to improve productivity but not about how to improve pay rates, working conditions, or seniority rules.

As in the quote by Matsushita cited above, the goal for many managers is to define problems in their own terms and to maintain the control that Taylor sought, while gathering the knowledge of those on the shop floor. W. Edward Deming, one of the main gurus of the "Quality" movement says that workers' ideas are important but that decisionmaking must remain with management. In an article on the role of unions, two of Deming's disciples recently wrote that, first and foremost, a union must "absorb and live the company's mission, goals and operating philosophy."[20]

Parents and day-care workers will recognize this brave new world of control. Smart parents know that if they want a child to sit in a chair, they should not order the child to do so but instead should ask, "Which chair would you like to sit in?" Having picked a chair, the child feels that he or she has made a free choice and has expressed his or her needs and interests. But the reality is that the parents have achieved their goal of having the child sit down.

Most worker participation programs actually give employees little access to information, little power to evaluate performance, and little ability to define the problem. In practically every case, final control over technological decisionmaking resides firmly in the hands of management. Even when a union is present, the fact of ultimate control by management is embedded in the contract. The following "management rights clause" is typical:

> Subject only to the limitations contained in this agreement, the Company retains the right to manage its business including (but not limited to) the right to determine the methods and means by which its operations are to be carried on, to assign and direct the work force, and to conduct its operations in a safe and effective manner.[21]

This bias toward giving management exclusive power over technology is reinforced by labor laws that do not make technology bargaining mandatory and by the ideology and culture of private property rights.

Even those with a generally progressive agenda, such as the professionals in this country who support "participatory design" of technology, need to be made acutely aware of the dangerous ground they are on when they fail to recognize the issues of power and control that lurk behind participation activities. Participatory design without a union, that is, without an independent, collective voice for the workforce, is an oxymoron.

What Can Be Done?

Once one sees that the core problem is the imbalance of power between management and labor, the collective strengthening of workers becomes *the* key answer. A collective voice is the only force that can stand up for the needs, interests, and hopes of the workers, especially in the face of rapid technological change. A requirement to democratize decisions concerning technology, therefore, is to strengthen labor unions. The much touted UTOPIA project in Sweden, in which workers and computer scientists designed a graphic work station for newspaper layout that used (rather than displaced) the traditional skills of typographers, occurred because of the strong position of trade unions in Scandinavia. "Experts" like computer scientists and engineers who want to make technology more responsive to the needs and interests of workers need to support and unite with unions.

The federal government also could play a positive role in the development of workplace technologies. Today, Washington primarily promotes technologies that serve the interests of the military and business managers, and any assistance for the needs or interests of workers occurs in the form of trickle-down economics. There are at least seven things the federal government could do that would support a move toward human-centered, worker-sensitive technology:

First, Congress should enact changes in the National Labor Relations Act (NLRA) that would make technology bargaining mandatory. This would go a long way toward changing the balance of forces in the battle over technology. In both Sweden and Germany management consults with workers through unions or through elected workplace councils *before* new technology is introduced.

Second, the government should strengthen laws that support unionization. Organizing a union under current law is very difficult, because management can resort to numerous delay tactics and because workers who want to unionize have minimal protection against retaliation from management. Even if over half the workers in a workplace sign cards saying they want to be in a union, an election can be put off for months or even years, giving the company plenty of time to "convince" (or threaten) the workforce not to unionize, to fire employees who are doing the organizing, and to wipe out the resources of the organizers. One out of twenty workers who votes for a union in this country gets fired. In addition, Congress should pass laws banning permanent replacements (anti-scab legislation) to help level the playing field.

Third, policies that promote economic security for the workforce are important mechanisms for supporting workforce involvement in union organizing, political change, and technology development. National policies are needed to raise unemployment benefits, to increase minimum wages and wage enforcement, to improve pay equity, to provide universal health care, and to permit workers to carry their benefits from job to job. These same measures would make it easier for employees to "vote with their feet" and to leave bad jobs for good ones, all of which would further strengthen the demand for worker-centered technologies and job design.

Fourth, the government should support broad worker-centered training. After World War II the GI Bill created a whole generation of skilled craftspeople, and estimates suggest that it paid for itself in increased tax revenues three to eight times over.[22]

Fifth, the government could underwrite more studies, performed by workers and trade-unions, on how technology is affecting the workforce and what worker-friendly alternatives are possible. The Ontario provincial government in Canada recently funded the Technology Adjustment Research Program (TARP), which supports projects run by over a dozen different unions.

Sixth, the government could put money into the development of worker-centered technologies. Instead of pursuing the workerless factory and artificial intelligence to eliminate human intervention, the government should be developing software that allows shop-floor workers to program computer-based equipment, to apply their vast knowledge to the operation of advanced equipment, and to make their jobs an integral part of the organization's learning process.

The government must try to reform the educational institutions that teach Taylorism to engineers and managers. If technologies are ever to be used and developed in ways that respect workers, the roots of Taylorism must be destroyed.

There are other things the government should do over the long run. It should work to eliminate all remaining vestiges of Taylorism. There should be a ban on short cycle-time work. Decisionmaking should be moved onto the shop floor and made part of everyone's job. Most fundamentally, democracy must come to the workplace.

For the sake of America's workers, and for the sake of our society as a whole, we need to develop new mechanisms for technological change. Technology must serve the needs of the many rather than the desires of just a privileged few. It must become a tool for building a more equitable society rather than a weapon for exacerbating inequalities. If technology is to serve the good of society, the key battles will have to be fought in America's workplace—and won by America's workers.

Notes

1. Frank Barkley Copley, *Frederick Taylor: Father of Scientific Management, Volume I* (New York: Harper and Brothers, 1923; repr. New York City: Augustus M. Kelley, 1969), p. 84, emphasis in original.
2. Harry Braverman, *Labor and Monopoly Capital,* (New York, NY: Monthly Review Press, 1974), p.118.
3. Copley, *Frederick Taylor,* p. 118.
4. Braverman, *Monopoly Capital,* p. 348.
5. Frederick W. Taylor, "Shop Management," in *Scientific Management* (Westport, CT: Greenwood Press, 1972), p. 187.
6. Ralph Barnes *Motion and Time Study: Design and Measurement of Work,* 7th ed. (New York: John Wiley and Sons, 1980), p. 16.
7. See Robert Karasek and Tores Theorell, *Healthy Work: Stress, Productivity, and the Reconstruction of Working Life* (New York: Basic Books, 1990).
8. Paul Kenneth Wright and David A Bourne, *Manufacturing Intelligence* (Reading, MA: Addison-Wesley Publishing, 1990), p. 6.
9. Wright and Bourne, *Manufacturing Intelligence,* p. 6.
10. Advertisement, *Apex Product News* (Dearborn, MI. Society for Manufacturing Engineers, 1989).
11. See Office of Technology Assessment, *Worker Training: Competing in the New International Economy* (Government Printing Office, Washington, DC: September, 1990); and National Center on Education and the Economy, *America's Choice: High Skills or Low Wages!* (Rochester, NY: June 1990).

12. Office of Technology Assessment, *Worker Training*; and National Center on Education and the Economy, *America's Choice*.
13. Office of Technology Assessment, *Worker Training*.
14. Anthony P. Carnevale, Leila J. Gainer, and Ann S. Meltzer, *Workplace Basics: The Skills Employers Want* (Alexandria, VA: American Society for Training and Development, United States Department of Labor Employment and Training Administration, 1988), p. 3.
15. Konosuke Matsushita, quoted in editor's note, *Manufacturing Engineering* February 1988, p. 15.
16. See Mike Parker and Jane Slaughter, Jane, *Choosing Sides: Unions and the Team Concept* (Boston, MA: South End Press, 1988).
17. "The Triple Revolution," 22 March 1964, reprinted in this volume.
18. *Ibid.*
19. *Ibid.*
20. Howard Seth Gitlow and Shelley Jann Gitlow, *The Deming Guide to Quality and Competitive Position* (Englewood Cliffs, NJ: Prentice-Hall, 1987), p. 208.
21. From the November 1989 contract between IBEW 2222 and the New England Telephone Company.
22. See *G.I. Course Approvals: A Report Prepared by the National Academy of Public Administration Foundation for the Veteran's Administration submitted to the Committee on Foreign Affairs* (Washington, DC: U.S. Government Printing Office, 1979), p. 22.

Chapter 4

The Conversation We Haven't Had: Technology, Trauma, and the Wild

by Chellis Glendinning

That millions of people share the same forms of mental pathology does not make these people sane.

—Erich Fromm

I met with a young political activist for conversation last week at my favorite cafe. A pro-feminist man and founder of an anti-war youth organization during the Gulf War, this 21-year-old lives to explore social issues and act on his convictions. His burning question of the hour concerned technology. "Has television made people less intelligent?" he wondered, and he based his conclusion on the deconstructionist dictate that one speak *only* from personal experience. His answer, "Decidedly not," and indeed, this young man's mental capacity was as substantial and his wit as clever as any I had seen of any age. But I could not help but notice that even before a quadruple espresso *latte* had exploded onto his brain cells, my young friend was ranting at 120-words-per-minute, vibrating in his seat like a rocket poised for take-off, and hurling about words like VPL, CDTV, HyperCard, and Macromind. And answering his own questions in quantam leaps across paradigms unintegrated by coherent worldview, physical reality, or moral obligation to life.

Like my friend, most of us who inhabit mass technological society find it difficult to understand technology's impact on social reality, let alone on our psyches. Like the tiny aerobic bacteria that reside within computer hardware, we are so entrenched in our technological world that we hardly know it exists. Yet widespread radioactive contamination, cancer epidemics, oil spills, toxic leaks, environmental illness, ozone holes, poisoned aquifers, and cultural and biological extinctions indicate that the technological construct encasing our every experience, perception, and political act stands in dire need of criticism. Further, such a critique requires integration by a coherent worldview, physical reality, and a moral obligation to life.

After we gaze at ourselves from outside our hardware, few of us will be able to deny that we need to reconstruct both our world and our worldview. Yet not many of us have begun the conversation that would lay a conceptual framework for such a shift. Instead, we remain trapped within both our hardware and our hardware-determined worldview, and as long as we remain so, technological development promises to march onward toward nanotechnology, virtual reality, genetic engineering, biological weapons, and all kinds of new technologies that will further disrupt ecological balance, cultural diversity, and psychological well-being.

I count myself as both technology critic and activist, proudly following in the footsteps of such insightful thinkers as philosopher Lewis Mumford, sociologist Jacques Ellul, historian David Noble, political scientist Langdon Winner, and social critic Jerry Mander.[1] The system under observation is mass technological civilization. As philosopher Martin Heidegger wrote, "We are questioning concerning technology in order to bring light to our relationship to its essence."[2] What *is* the essence of modern technology? How does it structure our lives? Our perceptions? Our politics? How does it shape our psyches? What does it say about our relationship to our humanness and to the Earth? Unfortunately, obstacles to answers are entrenched, like concrete piers at a freeway exchange, in both our social and psychological reality.

I discovered the scope of such obstacles while I was on a promotional tour for my book *When Technology Wounds*.[3] The book is based on a psychological study of technology survivors: people who have become medically ill as a result of exposure to some health-threatening technology. I interviewed Love Canal residents, atomic veterans, asbestos workers, DES daughters, electronics plant workers, Dalkon Shield survivors, homeowners whose groundwater had been contaminated, Nevada Test Site downwinders, sufferers of cancer, environmental illness, chronic fatigue immune dysfunction, and many other survivors.

By all accounts, this population is on the rise. Forty-one thousand Louisiana residents are exposed to 3.5 million tons of toxic landfill along the industrial corridor between Baton Rouge and New Orleans.[4] Thirty million U.S. households, or 96 million people, live within fifty miles of a nuclear power plant.[5] One hundred and thirty five million residents in 122 cities and counties breath consistently polluted air,[6] while 250 million Americans—everyone of us—are exposed to 2.6 billion pounds of pesticides each year, in addition to all the radioactive fallout ringing the globe from Hiroshima, Chernobyl, and the nuclear test sites in Nevada and Khazakstan.[7]

On the book tour I suggested that since people everywhere are getting sick from technological exposure, we had best enter into an informed and reasoned conversation about technology. Such a conversation was not forthcoming. In a debate on National Public Radio with MIT Professor Marvin Minsky, the founder of artificial intelligence, I was asked if I had any objections to computers. I expressed concern that the deadly chemicals used to manufacture computers such as chlorofluorocarbons, diethylamine, lencast, 4,4 isophopylidenediphenol, epichlorohydrin resin, and biosphenol A contaminate the biosphere. I mentioned Yolanda Lozano, a 36-year-old worker from a GTE plant in Albuquerque who died of cancer from chemical exposure on the job. In reply, Professor Minsky quipped, "It doesn't matter."[8] Elsewhere on my tour, the conversation ended before it began. "Get this woman off the air! She's the stupidest guest you've ever had!" shrieked one talk show listener. "I can't give up my mammogram!" howled another. "As soon as we take care of this environmental thing," insisted one man at a book fair, "we've got to colonize Mars. It's *imperative* for our belief in the future."[9]

Techno-Addiction

As a psychologist, I compare today's public awareness of the impacts of technology to that of alcoholism in the 1950s. Back then, everybody drank. It was more than socially acceptable to drink; it was required. Alcoholics Anonymous (AA) was twenty years old and growing, but still considered an embarrassment to its members. I recall how my mother placed a copy of the AA book on our coffee table in 1953, hoping against hope that my alcoholic father would pick it up. He never did.

It is not a new idea that we who live in mass technological society suffer psychological addiction to specific machines like cars, telephones, and computers, and even to technology itself. But the picture is bigger and more complex. As social philosopher Morris Berman says:

> Addiction, in one form or another, characterizes every aspect of industrial society. . . . Dependence on alcohol (food, drugs, tobacco . . .) is not formally different from dependence on prestige, career achievement, world influence, wealth, the need to build more ingenious bombs, or the need to exercise control over everything.[10]

The editor of *Science* magazine describes the nation's dependence on oil as an addiction,[11] while Vice President Al Gore claims that we are addicted to the consumption of the Earth itself.[12] Evolutionary philosopher Gregory Bateson points out that addictive behavior is consistent with the western approach to life that pits mind against body and concludes, "It is doubtful whether a species having *both* an advanced technology *and* this strange [polarized] way of looking at its world can survive."[13]

To clarify this notion that contemporary society itself is based on addiction, what I call "techno-addiction," we will do well to turn to the pioneers of the sociology of technology. Mumford, Ellul, Winner, and Noble remind us that no machine stands alone.[14] In other words, we will forever be trapped in a narcissistic "but-I-want-my-mammogram" analysis as long as we view technology only as specific machines that either serve us individually or do not. What Mumford calls the "mechanical order" or the "megamachine" is an entire psycho-socio-economic system that includes all the machines in our midst; plus all the organizations and methods that make those machines possible; plus those of us who inhabit this technological construct; plus the ways in which we are socialized and required to participate in the system; and the ways we think, perceive, and feel as we attempt to survive within it.

What I am describing is a human-constructed, technology-centered social system built on principles of standardization, efficiency, linearity, and fragmentation—like an assembly line that fulfills production quotas, but cares nothing for the people who operate it. Within this system, technology influences society. The automotive industry completely reorganized American society in the 20th century. Likewise, nuclear weapons define global politics. At the same time, society reflects the technological ethos. As Mander shows, the social organization of workplaces, as well as their architecture and physical layout, reflect the mechanistic principles of standarization, efficiency, and production quotas.[15]

From our everyday experience within mass technological society, we will note that "normal" acts like standing in line, obeying traffic signals, or registering for the draft all constitute acts of participation in this grand machine. Regarding our minds and bodies as disconnected in health and disease, or thinking that radioactive waste buried in the Earth won't eventually seep into the water table, are symptoms of the fragmented thinking that emerges from such a mechanical order.

At this point in history, technology and society are completely interwoven. The feminist philosopher Susan Griffin characterizes the impossible task of distinguishing the two as "like saying my hand causes my fingers."[16] "Technology has become our environment as well as our ideology," writes the Dutch social critic Michiel Schwarz, "We no longer use technology, we live it."[17]

Vine Deloria, a Sioux Indian and author of many books on Indian history and politics, describes the results of this social-technological imbrication as "the artificial universe."

> Wilderness transformed into city streets, subways, giant buildings, and factories resulted in the complete substitution of the real world for the artificial world of the urban man . . . . Surrounded by an artificial universe when the warning signals are not the shape of the sky, the cry of the animals, the changing of seasons, but the simple flashing of the traffic light and the wail of the ambulance and police car, urban people have no idea what the natural universe is like. . . . Their progress is defined solely in terms of convenience within the artificial technological universe with which they are familiar.[18]

Langdon Winner moves the idea further, arguing that the artifacts and methods invented since the technological revolution have developed in size and complexity to the point of canceling our very ability to grasp their impact upon us. The socially structured scientific-technological reality that now threatens to determine every aspect of our lives and encase the entire planet is out of control, he asserts. "Our technologies are tools without handles."[19]

Total immersion, loss of perspective, and loss of control tip us off to the link between the psychological process of addiction and the technological system. According to Craig Nakken of the Rutgers School of Alcohol Studies, addiction is a progressive disease that begins with inner psychological changes, leads to changes in perception, behavior, and life-style, and then to total breakdown.[20] The hallmark of this process is the out of control, often aimless compulsion to fill a lost sense of meaning and connectedness with substances like alcohol or experiences like fame.

Throughout the technological system, the recognized symptoms of the addictive process are blatantly evident. They are obvious in the behavior of those Jacques Ellul calls "the cabal of self-serving officials and executives"[21] who promote technology to maintain control over society or to inflate their own bank accounts and egos. And they are evident for us all because our experience, knowledge, and sense of reality have been shaped by life in the technological world. Some symptoms of the addictive process include denial, control, dishonesty, thinking disorders, grandiosity, and an unhealthy relationship with one's feelings.

Denial

According to psychotherapist Terry Kellogg, addiction is "a process of decreasing choice sustained by denial."[22] The practicing alcoholic pretends that everything is normal and holds up appearances at all costs. I once met a politician who had gotten himself elected to office by spending ten times more on his campaign than any other contender in the race. He was addicted to power, sex, work, overspending, and abuse of other people. His denial of these addictions blinded him from admitting them for years—until one day the Sheriff's Family Violence Protection Act served him notice for assaulting an old girlfriend.

The denial of addiction which we find in society's ecological, economic, and psychological crises resembles this man's life. A society-wide stance of "business-as-usual" pervades our lives. Denial abounds. The automotive industry at home and abroad keeps cranking out new models of polluting cars. Television runs ads for them. We continue to buy them. The U.S. government denies a link between technological development and global warming, while President George Bush calls for more technological development as the answer to environmental disaster. The plastics industry inundates world markets with petroproducts, even using the idea of park benches made from recycled plastic as an excuse for further production. The medical establishment denies the existence of environmental illness. And corporations deny the environmental impact of toxic manufacturing processes.

Technology survivors suffer further pain as they encounter widespread denial that their illnesses are caused by technology from the insurance industry, justice system, medical establishment, media, and even friends and family. As Love Canal resident Lois Gibbs told me:

> I went to my son's pediatrician, and I said, "Look, there are eight patients who have you as their doctor. All of them are under the age of twelve, all of them have a similar urinary disorder. Why is this? What do you make of the fact that you have eight patients who live within a few blocks of Love Canal who have *the same disease*?" He said, "There is no connection."[23]

Dishonesty

This symptom is acted out by the alcoholic in secret drinking, sneaky behavior, and lying about feelings and activities. With respect to technology addiction, dishonesty reveals itself most blatantly in the behavior of corporations and government agencies whose self-interest lies in purveying offending technologies. We know, for example, that officials at A.H.

Robins, the makers of the Dalkon Shield, knew in advance of the potential medical risk of their product. Nonetheless, they sent it to market, and when reports and studies indicating ill effects became public knowledge, A.H. Robins claimed complete ignorance.[24] Likewise, in the face of increasing liability suits for asbestos production, Manville Corporation (formerly Johns-Manville] froze its assets and insulated itself from lawsuits by declaring bankruptcy. At the time, Manville ranked 181st on the Fortune 500 list, with assets over $2 billion.[25]

Control

Addicts need to control their world to enjoy uninterrupted access to the source of their obsession. A workaholic I know who directs a small institute is incapable of negotiating even the smallest agreement, because input from others upsets her sense of control. Likewise, today's multinationals display an unbounded obsession with controlling the world's resources, consumer markets, workers' behavior, and public opinion toward their products. The Chinese government is similarly obsessed with controlling its citizens, resources, and public image, as were the pre-1989 regimes in Eastern Europe.

Let us also consider the very structure of modern technology. The kinds of technologies a society develops are not as absolute or preordained as our ethos of linear progress would have us believe; they express a society's goals, both conscious and unconscious. In mass technological society there exists a striking resemblance between the kinds of technologies produced and tyrannical modes of political power. Winner discusses the similarity in language between the realms of power and technology. "Master" and "slave" are words used to characterize both technology and fascist politics, while "machine," "power," and "control" appear in the vocabularies of both worlds.[26]

On a phenomenological level, we observe that when humans assume a position of extreme dependence on technical artifacts, the lines blur between who is master and who is slave. What happens to our lives when cars break down or telephones go out? What happens when you don't own a FAX machine or a computer or a car? Technology's mastery over our lives translates into political disempowerment as well. The very conception, invention, development, and deployment of new technologies involves a highly undemocratic social process that is rationalized as "progress."[27] The life experience of technology survivors attests to this fact: they are usually exposed to technological events that rob them of their health and livelihood without any warning or choice in the matter.

If the particular kinds of technologies in our midst exist to promote mastery and power, we might ask for whom? And over whom? Windmills and teepees express democratic and ecological values because the very people who invent, produce, and maintain them are the same people who use them. By contrast, the technologies disseminated in mass society reflect a mentality of control over the natural world, space, other people, and even ourselves. As Mander points out, running a nuclear power plant requires tight, centralized control by both government and industry first to produce such a capital-intensive project, then to master public opinion, and finally to provide military back-up in case of sabotage, accidents, or public protest.[28] The presence of nuclear, biological, and chemical weapons in a nation's arsenal not only controls that nation's enemies; it also frightens and intimidates, and thereby controls that nation's own citizens.

Thinking Disorders

Alcoholics are notoriously over-reliant on linear, rationalistic, either/or logic. Their thinking is typically obsessive, confused, and narcissistic. An alcoholic man who attacked a woman later explained that he was not responsible for her resulting medical bills; instead, he blamed this woman for her adverse reaction to his attack, rationalizing "you create your own reality."

Likewise, much thinking in mass technological society is dysfunctional. Many people embrace the "technological fix" as the answer to social, psychological, and medical problems caused by previous technological fixes. The parallel between this aspect of techno-addiction and the alcoholic who drinks to alleviate the physical and emotional pain of hangover is conspicuous. Doctors treat servicemen with cancer from exposure to nuclear testing with small blasts of radiation. Chemists look to ever more potent pesticides to conquer the insects resistant to last year's poison. A proposed government program seeks to cover the oceans with polystyrene chips which, it is hoped, will reflect "unwanted" sunlight off the Earth's surface and save us from global warming. Likewise, some scientists suggest orbiting hundreds of satellites around the planet to block the sun's light.[29] This is techno-addictive thinking at its most convoluted.

Grandiosity

The practicing alcoholic's delusion of inflated power is well known. The delusion of grandeur that fuels technological development is less apparent, more assumed. This grandiosity insists that mass technological society is superior to all other social arrangements. It implies that human evolution is linear and always progressive, and that all societies should be judged by the yardstick of technological achievement. These ideas are inescapably interwoven throughout our culture—on television, in textbooks, in movies, on news reports—and are reinforced by each season's parade of new, state-of-the-art technologies.

Technological society's main organ of socialization, public relations, purveys the grandiosity of technology. "Master the Possibilities," teases the MasterCard ad. "What Exactly Can the World's Most Powerful and Expandable PC Do? Anything It Wants," promises the Compaq Deskpro. At the same time, the "smart weapons" unleashed on television during Desert Storm advertise that American technology, and America, are "Number One." Behind this all-too-earnest insistence lies the out-of-control, often aimless compulsion to create ever-increasing expressions of grandiosity—and the hallmark of the addict—to return continually to the source of aggrandizement. We *need* more cars, more bombs, more delivery systems, more televisions, more golf courses, more dams, more shopping malls, more new technologies to prove our grandiosity.

Unhealthy Relationship with Feelings

Alcoholics are brimming with emotions, but they can't express themselves directly or constructively. Instead, their feelings are hidden from view in the shadows of their unconscious minds, and so they deny their feelings and live in a state of frozen emotion. Or they act out their feelings in the dishonest, controlling, grandiose behavior of the addict.

Likewise, survival in the technological system requires that we act "cool" and behave like machines. The hallmark of technological education is to learn mathematics to quantify reality and to master fragmented thinking to function in a mechanistic world. Every subject we learn in schools seems unrelated to the others. Each department of government appears disconnected from every other. Modern medicine denies links between body organs and systems; between mind, body, and society.

Similarly, mass technological society is structured "top-down," its fragmented nature keeping most of us from ever grasping an understanding of the whole. The Manhattan Project that built the bombs that killed hundreds of thousands of people in Hiroshima and Nagasaki was constructed according to a mechanistic military model. The Project included 37 installations scattered across the United States and Canada,[30] each providing one fragment of the production process. At the Los Alamos Laboratory, work was purposefully accomplished with a compartmentalization of tasks and a censuring of communication between scientists that enabled everyone involved to lose his or her sense of vulnerability and to engage in activities the consequence of which could neither be felt nor understood.

The upshot of such an approach to life is that feelings, experiences, and perceptions become disconnected from each other, and the unconscious mind becomes the receptor of repressed feelings. As a result, many of us tend to reside in a semi-conscious state: the hideous and subterranean violations around us catalyze our feelings, but unacknowledged and unwelcome by the mechanistic world, we act them out in behaviors we neither feel nor understand. Like dropping the atomic bomb.

Technology, Trauma, and the Wild

We must recognize systemic addiction in mass technological society if we are ever to achieve a state of psychological and technological well-being. The twelve-step recovery movement says that the addict must make "a searching and fearless moral inventory" of him or herself.[31] On the personal level, this undertaking includes claiming responsibility for instances in which we have violated another person's integrity. On the collective level, we would claim responsibility for technological society's uncounted violations against humanity, animals, the plant world, and the Earth. But lest our bleeding hearts overtake the process, let us be alert. As Kellogg tells us, addictive behavior is not natural to the human species. It occurs because some untenable violation has "happened *to* us."[32]

And indeed, we have undergone an untenable violation: a collective *trauma* that explains the insidious reality of addiction and abuse infusing our lives in mass technological society. *The Diagnostic and Statistical Manual of Mental Disorders* defines trauma as "an event that is outside the range of human experience and that would be markedly distressing to almost anyone."[33] Psychiatrist Abraham Kardiner describes it as "an external influence necessitating an abrupt change in adaptation which the organism fails to meet."[34] The trauma endured by technological people like ourselves is the systemic and systematic removal of our lives from the natural world: from the tendrils of earthy textures, from the rhythms of sun and moon, from the spirits of the bears and trees, from the life force itself. This is also the systemic and systematic removal of our lives from the kinds of social and cultural experiences our ancestors assumed when they lived in rhythm with the natural world.

Deloria rightfully infers that we technological people "have no idea" about much of anything that resides outside "the artificial technological universe with which [we] are familiar." Human beings evolved over the course of some three million years and 100,000 generations in synchronistic evolution with the natural world. We are creatures who grew from the Earth, who are physically and psychologically built to thrive in intimacy with the Earth. A mere 300 generations ago, or .003 percent of our time on Earth, humans in the western world began the process of controlling the natural world through agriculture and animal domestication. Just five or six generations have passed since the industrial societies emerged out of this domestication process. Our experience in mass technological society is indeed "outside the range of human experience," and by the evidence of psychological distress, ecological destruction, and technological control, this way of life has necessitated "an abrupt change in adaptation."

Though largely ignored, evidence jumps from the pages of anthropological texts suggesting that the very psychological qualities so earnestly sought in today's recovery, psychological, and spiritual movements, and the very social equalities for which today's social justice movements struggle valiantly, and the very ecological gains sought after by today's environmental movements all comprise the same qualities and conditions in which our species lived for over 99.997 percent of its existence.

For nature-based people, addiction was not a normal occurrence gripping members of society or negatively guiding decisions reached in council. Rather, people lived every day of their lives *in the wilderness*. We are only beginning to grasp how such a life served the inherent expectations of the human psyche for development to full maturation and health. In nature-based people who today maintain some vestiges of their relationship to the Earth and their Earth-based cultures, we can discern a decided sense of ease with daily life, a marked sense of self and dignity, a wisdom which most of us can admire only from afar, and a lack of the addiction and abuse that have become systemic in civilization.

Anthropologists report that in small, face-to-face communities most nature-based people of the past practiced an easy form of democracy. Every member had a say and members of the community listened to each other.[35] Workaholism did not exist. Rather, leisure time marked the nature-based experience, with the average work day lasting three to five hours.[36] The population tended to remain stable, held in check by natural (rather than technological) fertility-control factors arising out of the nature-based diet and lifestyle.[37] And ecological sustainability reigned: all tools were made of natural substances, and constant movement from one place to another allowed for remaining waste to biodegrade into the Earth.

As psychohistorian Paul Shepard notes:

> White, European-American, Western peoples are separated by many generations from decisions of councils of the whole, life with few possessions, highly developed initiation ceremonies, natural history as everyman's vocation, a total surround of non-man-made otherness with spiritual significance, and the 'natural' way of mother and infant.[38]

The loss of these psychological and cultural experiences in the face of an increasingly human-constructed and eventually technology-determined reality, and the loss of living in fluid participation with the wild, constitute the trauma we have inherited.

The hallmark of the traumatic response is dissociation. This is also the psychological result of the kinds of social changes that took place in the process of domestication. Shepard describes this process as the initiation of a heretofore unheard of tame/wild dichotomy in which all things considered tame (domesticated seedlings, captured animals, and the mechanical and controlling mentality required to keep them alive) are prized and protected, while all things considered wild ("weeds," wild animals, and the fluid, participatory way of being human) are considered threatening and to be kept at bay.[39]

This split between wild and tame lies at the foundation of both the addictive personality and technological society. Ultimately, such a split imprisons us in our human-constructed reality and causes all the unnecessary and troublesome dichotomies with which we grapple today—from male/female and mind/body, to secular/sacred and technological/Earth-based. Shepard writes:

> In the ideology of farming, wild things are enemies of the tame; the wild Other is not the context but the opponent of 'my' domain. Impulses, fears and dreams—the realm of the unconscious—no longer are represented by a community of wild things with which I can work out a meaningful relationship. The unconscious is driven deeper and away with the wilderness.[40]

In his work on the post-traumatic stress of individuals, psychiatrist Ivor Browne describes dissociation in terms remarkably similar to Shepard's. He sees it as "unexperienced experience,"[41] experience that has not been properly processed by the psyche and integrated into memory. Just as animals often meet threats of disaster by "playing dead," so traumatized people split their consciousness, "freezing" the experience of loss and pain from awareness and thus "playing dead" in mind and body. This mechanism is a brilliant way to protect the psyche from threats that our nervous systems simply were not built to handle, and changes that we were never meant to integrate.

The purpose of such dissociation, then, is self-preservation. As Ivor Browne describes it: "Whenever we are faced with an overwhelming experience that we sense as potentially disintegrating, we have the ability to suspend it and 'freeze' it in an unassimilated, inchoate form and maintain it in that state indefinitely, or for as long as necessary."[42] Technological people's dislocation from the only home we have ever known is a traumatic event that has occurred over generations, and that occurs again in each of our childhoods and in our daily lives. In the face of such a breach, symptoms of traumatic stress are no longer the rare event caused by a freak accident or battering weather, but the everyday stuff of everyman's and everywoman's life experience.

Terry Kellogg describes trauma as "the freeway to addiction,"[43] and his technological metaphor is not lost. As human life comes to be structured increasingly by mechanistic means, the psyche restructures itself to survive. The technological construct erodes primary sources of satisfaction once found routinely in life in the wilds, such as physical nourishment, vital community, fresh food, continuity between work and meaning, unhindered participation in life experiences, personal choices and community decisions, and spiritual connection with the natural world. Bereft and in shock, the psyche finds some temporary satisfaction in pursuing secondary sources like drugs, violence, sex, material possessions, and machines. While these stimulants may satisfy in the moment, they can never truly fulfill primary needs. And so the addictive process is born. We become obsessed with secondary sources as if our lives depended on them.

Today the world is awash in a sea of both personal and collective addictions: alcoholism, drug abuse, sex addiction, consumerism, eating disorders, codependence, warmaking, and global drama. Psychotherapist and author Anne Wilson Schaef points out that beneath these behaviors lies an identifiable disease process "whose assumptions, feelings, behaviors, and lack of spirit lead to a process of nonliving that is progressively death-oriented."[44] While her words describe the addictive process of individuals, they also characterize the techno-addiction of a civilization. Society is addicted to specific technologies like cars, supercomputers, and biological weapons, all of which facilitate an unhealthy propensity to control, numb the psyche from pain, and momentarily feed a craving for power.

Techno-addiction is also an addiction to a way of perceiving, experiencing, and thinking. As the world has become less organic and more dependent on techno-fixes for problems created by earlier techno-fixes, humans have substituted a new worldview for one once filled with clean rushing waters, coyotes, constellations of stars, tales of the ancestors, and people working together in sacred purpose. But the ancestors from the western world took on the crucial task of redefining their worldview in a state of psychic dislocation, and so they ended up projecting a worldview that reflects the rage, terror, and dissociation of the traumatized state. They dreamed a world not of which humans are fully part, but one that we can wholly define, compartmentalize, and control. They created linear perspective, the scientific-technological paradigm, and the mechanistic worldview.

Life on Earth encased in the product of such a construction is, to quote the Hopi, hopelessly *koyanaskatsi,* or out of balance. As a psychologist, I believe that to address this imbalance at its roots will require more than public policy, regulation, or legislation. It will require a collective psychological process to heal us technological peoples who, through a mechanized culture, have lost touch with our essential humanity.

The Conversation

Just as recovery programs for individuals initiate a process of healing by using existing pain to break through denial, so we in technological society must face the challenge of healing by first recognizing our pain. Our lives are fraught with pain. Whether we approach this suffering on the level of personal addictions, economic disparities between rich and poor, cancer epidemics, toxic dumping, violence against women, or the prospect of life's extinction, human pain lies at the groundlevel of any authentic conversation we might have about technology.

The first obstacle to such a conversation is psychological: the addict really does believe his or her own denial of addiction. Rationally explaining to Professor Minsky that Yolanda Lozano's death does matter lands on ears attuned only to the bleep of computer language. And my talk show listener does believe that I am an intellectual nincompoop. A second obstacle is worse. For all the enlightenment provided by the alcoholism/techno-addiction metaphor, it becomes a bit flimsy here. I am beginning to feel as I imagine Sigmund Freud might have felt writing *Civilization and Its Discontents* as Hitler's power grew: that the grave disorders residing in the human psyche are based in the construction of civilization itself, and that psychoanalysis can only, at best, offer palliative relief.

We must make some choices. Do we rally 'round the technological fix one more time, with hopes of busing the ozone from city streets (where it is too plentiful) to the stratosphere (where it is too thin)? Do we hunker down behind the psychic numbing and cynicism of our ever-blaring computer screens? Do we muster our faith and embark upon some form of collective recovery toward the possibility of aliveness and responsibility? Or do we attempt a combination of these three, and any others we can conjure up?

Because I am insufferably adamant about the need to throw off the chains (freeways, electromagnetic force fields, and nuclear weapons) of mass technological society and become as wild as the Earth intended, I am dedicated to recovery. This is recovery not just of the "unexperienced experience" brought on by the trauma, violation, and abuse we sustain, but of the joy, laughter, and compassion we so sorely miss. It is, as Morris Berman writes, "the recovery of our bodies, our archaic traditions, our unconscious mind, our rootedness in the land, our sense of community, and connectedness with one another."[45] The choice to engage in such a process of recovery is painful. Yet the acceptance of pain, rather than its avoidance, marks the first step in the de-mechanization of the technological psyche.

Though seemingly small and individualized, I want to offer for our consideration the inclusion of a psychological path: the recovery of a respectful intimacy with the natural world. Those who have made a transition in their relationship with the Earth from one of speeding over it at 75 miles per hour, throwing aluminum cans into its rivers, or simply not caring—to one of wonderment, curiosity and loyalty—will attest to the delight here and to the discovery of long-forgotten strengths and convictions. But because of the long-standing breach of intimacy we technological peoples have endured, there also is pain. There is pain in opening ourselves to a world being wantonly destroyed before our eyes. Radioactive fallout rings the globe. The trees of the Black Forest are withering. The seals in the North Sea are dying of immune-deficiency disease. And the poisons keep leaking out.

Then, there are our feelings about how the natural world for most of us is nonexistent in our lives today. We cannot see the moon overhead because of the smog. Our feet do not touch the Earth when we walk. We rarely walk. We spend our time fixated before electromagnetic screens. We do not know how to speak to or learn from the natural world.

Finally, we have feelings about how as children we were rarely encouraged or taught or given the context within which to establish an authentic relationship with the natural world. This latter relationship is essential to our growth into mature human beings, and was built into our lives throughout our evolution. Yet technological society denies us this intimacy from the start. As children, we barely learn how birds lay eggs, how the moon changes shape, how snow is cold, or how the Earth can answer our questions and help us. The excavation of feelings about this lost past, and about our present losses, provides the psychic crucible for a nonmechanistic way of being; as we feel, we come alive.

A second conversation I want to suggest is also fraught with pain. That is, to embrace the human face of our technologized world—to acknowledge the Yolanda Lozanos among us who, whether from AIDS, environmental illness, cancer, radioactive contamination, pesticide exposure, or air pollution, do not have the psychic luxury of ignoring the state of the planet. In conducting my research for *When Technology Wounds*, I first located interviewees by asking colleagues if they knew of any technology survivors. Everyone I asked knew someone who had been poisoned by dioxin, had become sterile from the Dalkon Shield, or suffered from chronic fatigue immune dysfunction, environmental illness, or cancer. The number of people who can now identify themselves as technology survivors grows every day. The task now is to let the awesome scope and wrenching reality of human-size technology-induced suffering into our hearts.

And "What will come of such conversations?" ask you who worry that heart-felt experience supplants political ambition. One overlooked outcome of any authentic recovery process is that it refutes personal powerlessness. My own experience tells me that because the clear-sightedness and love for life that stream forth are so mighty, the passion to become a political player can be held back by neither bulldozer nor virtual reality. The words that spring to mind when I am faced with yet another violation of life are these: "Over my dead body . . ." And when I see an opportunity to create something new: "To Life!"

Because we know ourselves to be made from this earth.
—Susan Griffin

Notes

1. Lewis Mumford, *The Myth of the Machine: Technics and Human Development* (New York: Harcourt Brace and World, 1967); *The Myth of the Machine: The Pentagon of Power* (New York: Harcourt Brace and World, 1970); *Technics and Civilization* (New York: Harcourt Brace, 1934); Jacque Ellul, *The Technological Society* (New York: Vintage, 1964); *The Technological Bluff* (Grand Rapids, MI: Eerdmans, 1990); David Noble, *America By Design* (New York: Alfred Knopf, 1977); Langdon Winner, *Autonomous Technology* (Cambridge, MA: MIT Press, 1977); *The Whale and the Reactor* (Chicago: University of Chicago Press, 1986); Jerry Mander, *Four Arguments for the Elimination of Television* (New York: Quill, 1978); and Jerry Mander *In the Absence of the Sacred* (San Francisco, Sierra Club Books, 1991).
2. Martin Heidegger, "The Question Concerning Technology," in *Technology and Other Essays*, trans. William Levitt (New York: Harper & Row, 1977), p. 3.
3. Chellis Glendinning, *When Technology Wounds* (New York: William Morrow, 1990).
4. David Maraniss and Michael Weisskoff, "Corridor of Death Along the Mississippi," *San Francisco Chronicle*, 31 January 1988; and Jay Gould, *Quality of Life in American Neighborhoods (Boulder, CO: Westview Press, 1986)*, pp.2.117-2.120.

5. Critical Mass Energy Project, "The 1986 Nuclear Power Safety Report" (Washington, DC: Public Citizen, 1986); and Daniel F. Ford, *Three Mile Island* (New York: Penguin, 1982).
6. *Aerometric Information and Retrieval System: 1988 with Supplemental Data from Regional Office Review* (Washington, DC: Environmental Protection Agency, July 1989).
7. *Unfinished Business: A Comparative Assessment of Environmental Problems* (Washington, DC: Environmental Protection Agency, Office of Policy Analysis, February 1987), pp. 84-86; Lawrie Mott and Karen Snyder, "Pesticide Alert," *Amicus Journal*, 10:2 (Spring 1988), p. 2; and *Information Disease Almanac, 1986* (Boston: Houghton Mifflin, 1986), p. 129.
8. "Neo-Luddism Is Sweeping North America: An Interview with Chellis Glendinning and Marvin Minsky," Canadian Broadcasting Company rebroadcast on National Public Radio, 26 March 1990.
9. "The Mike Cuthberg Show," WAMU-FM, 16 May 1990.
10. Morris Berman, *The Re-Enchantment of the World* (New York: Bantam, 1981), p. 242.
11. D.E. Koshland, "War and Science," *Science*, 251:4993 (February 1, 1991), p. 497.
12. Al Gore, *Earth in the Balance* (New York: Houghton Miflin, 1992).
13. Gregory Bateson, *Steps to an Ecology of Mind* (New York: Random House, 1972), pp. 309-337.
14. Mumford, *Technics and Civilization*, p. 12; *The Pentagon of Power*, Chapter 13; Ellul, *The Technological Society*, pp. 3-11; Winner, *Autonomous Technology*, pp. 8-12; and Noble, *America By Design*, Chapters 8-9.
15. Mander, *In the Absence of the Sacred*, Chapter 7.
16. Author's conversation with Susan Griffin, Berkeley CA, 15 June 1987.
17. Michiel Schwarz and Rein Jansma, eds., *The Technological Culture* (Amsterdam: De Bailie Publishers, 1989), p. 3.
18. Vine Deloria, *We Talk, You Listen* (New York: Delta, 1970), p. 185.
19. Winner, *Autonomous Technology*, p. 29.
20. Craig Nakken, *The Addictive Personality* (San Francisco: Harper & Row, 1988), pp. 19-62.
21. Jacques Ellul, *Propaganda: The Formation of Men's Attitudes* (New York: Vintage Books, 1965), p. 121.
22. Terry Kellogg, "Broken Toys, Broken Dreams," (Santa Fe, NM: Audio Awareness, 1991). Audiotape.
23. Glendinning, *When Technology Wounds*, p. 66.
24. Morton Mintz, *At Any Cost: Corporate Greed, Women and the Dalkon Shield* (New York: Pantheon Books, 1985), Chapter 3.
25. Paul Brodeur, *Outrageous Misconduct: The Asbestos Industry on Trial* (New York: Pantheon Books, 1985), Chapter 10
26. Winner, *Autonomous Technology*, p. 20.
27. Mander, *In the Absence of the Sacred*, Chapter 7.
28. Mander, *Four Arguments for the Elimination of Television*, p. 44.
29. Mander, *In the Absence of the Sacred*, p. 179.

30. Richard Hewlett and Oscar Anderson, Jr., *The New World, 1939-1946: A History of the Atomic Energy Commission* (University Park, PA: University of Pennsylvania, 1962), p. 3.
31. Anne Wilson Schaef, *Co-Dependence* (San Francisco: Harper and Row, 1986), p. 27.
32. Kellogg, "Broken Toys, Broken Dreams."
33. *Diagnostic and Statistical Manual of Mental Disorders*, 3rd ed. (Washington DC: American Psychiatric Association, 1987).
34. Abraham Kardiner, "The Traumatic Neurosis of War," *Psychomatic Monograph II-III* (New York: P. Hoeber, 1941), p. 181.
35. Peter Wilson, *The Domestication of the Human Species* (New Haven, CT: Yale University Press, 1988), pp. 42-43; Stanley Diamond, *In Search of the Primitive* (New Brunswick, NJ: Transaction Books, 1974), Chapter 8; and Colin Turnbull, *The Forest People: A Study of the Pygmies of the Congo* (New York: Anchor, 1962), Chapter 6.
36. Marshall Sahlins, *Stone Age Economics* (New York: Aldine De Gruyter, 1972), pp. 14-32; Frederick McCarthy and Margaret McArthur, "The Food Quest and the Time Factor in Aboriginal Economic Life," in C.P. Mantford, ed., *Records of the Australian-American Scientific Expedition to Arnhem Land*, Vol. 2, Anthropology and Nutrition (Melbourne, Australia: Melbourne University Press, 1960), pp. 92-192; and Richard Lee, "iKung Bushman: Subsisstence: An Input-Output," in A. Vayda, ed. *Environment and Cultural Behavior* (Garden City, NJ: Natural History Press, 1969), pp. 59-74.
37. Margaret Ehrenberg, *Women in Prehistory* (Norman, OK: University of Oklahoma Press, 1989), pp. 85-90; L. Binford, *An Archaeological Perspective* (New York: Seminar Press, 1972); and Mark Nathan Cohen, *Health and the Rise of Civilization* (New Haven, CT: Yale University Press, 1989), p. 109.
38. Paul Shepard, *Nature and Madness* (San Francisco: Sierra Club Books, 1982), p. 120.
39. *Ibid.*
40. *Ibid.*
41. Ivor Browne, "Psychological Trauma, or Unexperienced Experience," *Revision*, 12:4 (Spring 1990), p. 26.
42. Browne, "Psychological Trauma, or Unexperienced Experience," p. 27.
43. Kellogg, "Broken Toys, Broken Dreams."
44. Wilson Schaef, *Co-Dependence*, p. 21.
45. Berman, *Re-Enchantment*, p. 282.

Chapter 5

Ending Toxic Pollution through Environmental Democracy

by Sanford J. Lewis

There is an emerging consensus that environmental laws regulating toxic chemicals and hazardous wastes are woefully inadequate. In its 1992 book, *Changing Course,* the Business Council for Sustainable Development, comprised of business leaders from 50 major corporations, wrote:

> [The] overuse and misuse of resources is accompanied by the pollution of the atmosphere, water and soil—often with substances that persist for long periods. With a growing number of sources and forms of pollution, this process also appears to be accelerating . . . . Clearly action is required. But which actions, and when, given the huge uncertainties involved?[1]

The prevalent strategy to control pollution by setting standards for "acceptable" emissions of toxics is failing to protect public health and environment. Recently, a more far-reaching solution has been proposed—that industry should reduce its *use* of toxic substances.[2] So far, corporations and federal policymakers have responded by encouraging "voluntary" reductions. The rationale for this approach is that public awareness of toxic emissions, combined with the escalating costs of pollution control, are forcing corporations to implement use-reduction measures. These same corporations assert that imposing legal standards for use-reduction could deny them the flexibility to make sound business judgments about what materials to use and about which production methods to employ.

But do these rationales really justify the current emphasis on voluntary reductions of toxics use? What kinds of results are the voluntary efforts producing? What are the alternatives? This essay answers these questions by exploring the current status of toxic chemical problems, the shortcomings of the traditional emissions regulations, and the limits to existing federal laws that are supposed to create incentives for toxics-use reduction. Three strategies to promote voluntary corporate action for toxics reduction are analyzed: the Environmental Protection Agency's (EPA) Industrial Toxics Program, the Chemical Manufacturers Association's "Responsible Care" Program, and the Ceres Principles, a code of conduct advanced by environmentalists and "socially responsible" shareholder groups. Finally, an alternative framework is suggested that integrates the advantages of the voluntary programs with needed legal and regulatory changes.

The Problem: An Environmental Health Crisis

Since 1940, the annual production of synthetic organic chemicals in the Untied States has increased rapidly from 2.2 billion to 214 billion pounds. This has been accompanied by an enormous increase in chemical waste disposal. However, public awareness of the extent of toxic waste disposal and the concommitant health hazards has emerged only recently.

Two federal programs established during the 1980s have been pivotal in creating a dramatically new understanding of the magnitude of the chemical waste-disposal problem. Through the Superfund program, the EPA prepared an inventory of hazardous waste sites. As of 31 December 1990 the EPA identified 32,645 dangerous sites and, of these, put more than 1,200 on its National Priority List of the sites with the worst contamination.[3] These priority sites are slated for cleanup under EPA's oversight.

Through the Community Right to Know Act, the federal government required industries to disclose the levels of toxic chemicals they were discharging into the air and water and onto the land. These industries reported their estimated emissions through the Toxic Release Inventory (TRI), which the EPA then published and made available to the public. In 1989, according to the EPA, 22,650 industrial plants and sites across the United States released 5.78 billion pounds of toxic chemicals into the environment. Actual emissions are far higher, because the approximately 330 hazardous substances monitored by the Toxics Release Inventory do not include more than 500 toxic chemicals regulated under other environmental laws. What's more, many companies are flouting the Right to Know law and have not yet filed the legally required information.[4] Although the EPA has enforcement authority, it lacks sufficient personnel to ensure full compliance by each individual company. A more accurate estimate is that about one trillion pounds of toxic chemicals are being released into the environment in the United States each year, primarily from industrial facilities.[5,6]

These massive chemical releases have reached populated areas across the country. According to the EPA, at least 140 chemicals foreign to human beings are now embedded in the bodies of Americans. Today all adults in the United States have measurable quantities of styrene in their fatty tissues; 100 percent have ethyl phenol; 96 percent have ethyl benzene; 96 percent have chlorobenzene; 96 percent have benzene; 91 percent have toluene; 83 percent have PCBs; and 93 percent have DDE, a breakdown by-product of DDT.[7] If the breast milk from American mothers were bottled and sold as a commercial product, it would probably be banned by the U.S. Food and Drug Administration, because it is so contaminated with pesticides and industrial poisons.[8]

While there has been an explosion of data and analysis regarding the extent of human and environmental exposure to toxic substances, a precise scientific consensus about the health effects will take a long time to develop. The few conclusive studies of Superfund sites and other settings suggest that the dangers are severe. Tests of many of the chemicals on animals and epidemiological studies conducted in workplaces and in some communities reveal that the waste products of the chemical industry, once released into the water people drink and the air they breathe, can cause cancer and can attack virtually every organ system of the human body.

The United States Agency on Toxic Substances and Disease Registry charged the National Research Council (NRC) of the National Academy of Sciences with reviewing available data and estimating the extent of health problems at Superfund sites. In its 1991 report, the NRC was unable to say exactly what portion of the U.S. population had been harmed by hazardous wastes, but it pointed to various studies that identified populations with serious health impairments seemingly correlated with the toxic contamination of air, water, or soil.[9] For instance, a study of lung cancer in all U.S. counties found that excessive rates of cancer were associated with four manufacturing industries: paper, chemicals, petroleum, and transportation (where workers are exposed to solvents and paints).[10]

The NRC's conclusions reaffirmed concerns raised elsewhere. Scientists have long known the potential severity of chemical hazards from testing on animals and from epidemiological analyses conducted in workplaces.[11] Numerous studies have shown that chemists tend to die of cancer more frequently than the rest of the population. A recent study of Exxon employees revealed that scientists, engineers, and research technicians working in the company had a greater risk of leukemia and lymphatic cancers than did managerial employees with less chemical exposure in the workplace.[12]

In 1990 the National Cancer Institute reported a 28 percent increase in the incidence of childhood cancer from 1950 to 1987.[13] More than 20 studies in the United States and elsewhere have demonstrated clear correlations between childhood cancers and exposure to carcinogenic chemicals. The three most common childhood malignancies—kidney and brain cancers, and acute leukemia—are often related to the occupational exposure of fathers and mothers to organic solvents, hydrocarbons, lead, paints, dyes, pigments, and pesticides.[14]

Despite this pervasive evidence, some scientists still question whether releasing toxics into the environment is harmful. The majority of public health professionals agree, however, that the wisest course is to *reduce all unnecessary exposures.* This raises further questions about what dose levels should be permissible and about who should shoulder the costs, especially when the chemicals already have been released into the environment and must be cleaned up.

Toxic Emissions-Control Standards: The Traditional Approach

The nation's Superfund program is slowly cleaning up the toxic messes of the past. Even if all of the existing sites of toxic pollution were effectively cleaned up, however, emissions of toxics into the environment would continue from current industrial operations. The government's principal regulatory strategy to reduce these emissions has been to adopt and enforce standards for emissions and for waste-disposal practices. This is the approach that prevails today, with new standards coming into effect during the 1990s as a result of amendments to the Clean Air Act and Clean Water Act. Yet emissions reduction has proven to be a defective, cumbersome, and costly way of handling toxic pollution. At least seven problems have plagued this strategy:

There Is Often No Truly "Safe" Level of Emissions

The standards assume the existence of an emissions level that is safe, but the evidence increasingly suggests that there is no safe level. The whole notion of setting permissible standards for persistent toxic substances is now being called into question. For instance, the International Joint Commission, a joint agency of the U.S. and Canadian governments overseeing transboundary issues, issued a report in 1992 on water quality in the Great Lakes[15] concluding that persistent toxic substances—those that remain in the environment for a long time and bioaccumulate in plants and animals—are too dangerous to the biosphere and to humans to permit their release in any quantity. Many of these substances disrupt the glandular systems of organisms, which can compromise immune systems, cause gross birth defects, and make people vulnerable to cancer. Recent studies also have shown that some of the substances are likely to effect the hormonal balance of men and women. Even if discharged in very small quantities, toxics stay in the environment and concentrate as they move up the food chain. That is, they exist at low levels in plants, at higher levels in fish that eat the plants, and at still higher levels in people who eat the fish. While the EPA may allow ten parts per billion of a substance in the environment today, if the chemical persists and accumulates in the food chain, as many chlorinated compounds do, it ultimately may cause more damage than the low levels of emissions might indicate.

Enforcement Is Weak

Environmental and occupational health regulators must rely on industries themselves to monitor and report pollution. Many small- to medium-sized hazardous waste generators are visited by environmental officials as infrequently as once every ten years. Occupational safety and health inspections can be even less frequent. According to a study by the AFL-CIO,[16] "Workplace inspections by OSHA have become a once-in-a-lifetime experience. OSHA can now inspect workplaces on the average only once every 84 years. And surprise inspections of high-hazard workplaces occur only once every 25 years. Even the best states would need more than ten years at current rates for its [sic] inspectors to get to each job site. Nevada, at eleven years, rates the best." With little threat of a costly enforcement penalty being imposed, it often pays for a company to pollute until their violation is detected. The standards represent an ideal, but not the reality, of how companies handle their toxics.

Pollutants Are Continuously Shifted to the Least Regulated Environmental Medium

As standards become stringent for one environmental medium, industries respond by dumping toxics into others. For instance, if a company finds that air pollution standards for one chemical have become too stringent, it may install scrubbers (that wash the chemical from the smokestack) and dump the liquid waste into a local stream. Regulators play a constant game of catch-up to find the path that toxics are taking and to head them off. Standards encourage firms to install emission controls rather than to scrutinize their use of toxics.

Continued Use of Toxics Affects Some Parts of the Population More than Others

Occupational health standards generally allow workforce exposures at ten times the levels permissible for environmental exposures. At least

100,000 workers die each year in the United States from job-related diseases, and more than 400,000 contract these diseases annually.[17] Besides the chemical industry itself, the manufacturing and service sectors often expose workers to considerable doses of toxics. Every year farm workers apply over one billion pounds of pesticides, suffer up to 300,000 related illnesses, and have miscarriages at seven times the national rate.[18] In addition, since the existing strategy encourages the siting of waste-disposal facilities in poor neighborhoods, it places an inequitable toxic burden on lower-income and minority communities. These communities receive a disproportionately higher level of toxic dumping and emissions because of their relative powerlessness to block siting permits.[19]

Accidents Will Happen

As long as industries continue employing highly dangerous chemicals, an American Bhopal is becoming increasingly possible as industrial facilities and their equipment age. The United States already has had many serious accidents. For instance, the EPA counted 6,928 accidents involving toxic chemicals between 1980 and 1985 that killed 135 and injured nearly 1,500 people. These numbers fail to indicate the full extent of the havoc that these accidents cause, including the evacuation of thousands from their homes.

Indoor Air Pollution Is Increased

Toxic substances find their way into homes and offices through cleaning products, roach sprays, termite poisons, paints, carpet backings, and so forth. We are just beginning to learn about the health effects of indoor air pollution—and the news is not good. In our relatively air-tight homes, the concentrations of pollutants in consumer goods can quickly accumulate to ten or a hundred times the levels regarded as safe.

Economic Inefficiency of Uniform Pollution-Control Strategies

Some authors have criticized emissions-control strategies from an economic viewpoint, asserting that the present system "wastes tens of billions of dollars every year, misdirects resources, stifles innovation, and spawns massive and often counterproductive litigation."[20] By inducing industries to buy the "state-of-the-art" pollution-control equipment, current policies do not prevent pollution at the point of production, where it might be least costly to do so. Industries know better than the government where these improvements can be made. The adversarial system of emissions controls encourages industries to invest in litigation rather than in pollution control. It also discourages investment in new technologies that might be developed if incentives were structured differently.

Toxics-Use Reduction

An alternative to the emissions-focused approach is to reduce the use of toxic chemicals and the production of hazardous wastes. Unlike the "end-of-pipe" pollution controls encouraged by emissions standards, this approach requires changing production methods, processes, and products. Five methods of reducing reliance on toxics have been widely discussed.

- *Substituting toxic chemicals used in production processes:* A company can substitute non-toxic or less toxic materials for highly toxic inputs. Some companies, for example, have begun substituting water- or citrus-based cleaning compounds for chemical solutions.
- *Changing the end product:* A company can change the design of a product so that production requires fewer toxics inputs. Some companies have found that they can strip paint from surfaces using mechanical processes similar to sandblasting rather than using toxic cleaners like turpentine.
- *Modifying or modernizing the production line:* A company can make fundamental process changes by replacing or changing production equipment, technologies, and methods. A change in the design of a production line often can eliminate the need to use a toxic chemical.
- *Better housekeeping:* With improved monitoring and closer attention to everyday practices of cleanup and maintenance, a company can better eliminate toxic "leaks" throughout the firm. For instance, instead of allowing chemicals to seep out, drop by drop, from a process vessel, a hole can be plugged or an aging valve replaced.
- *Recycling materials within the production process:* A company can reduce its waste streams by creating "closed loops" within processes that re-use certain potential hazardous wastes. This can lower the quantity of toxic and hazardous wastes that leave the facility or contaminate the work environment.

These approaches to "toxics reduction" apply to more than "hazardous wastes." They also apply to preventing toxic releases into the air through smokestacks, into the water through discharge pipes, and onto the land through dumping. They can reduce risks to the environment and to the health of workers and consumers without shifting risks *between* the environment, workers, or consumers. To summarize: The best way to reduce toxic waste and emissions is to lower the use of toxic chemicals.

Toxics-use reduction also may be as good for the economy as for the environment. It can improve the profitability of businesses far more than capital expenditures geared solely toward "end-of-the-pipe" emissions controls. An industrial process that uses toxic chemicals creates diverse liabilities and hidden costs for both firms and society. Conversely, an audit that reveals ways to reduce the use of toxics can help cut hidden costs and thereby improve business profits. One experienced consultant says that he can virtually guarantee that his waste reduction audits will pay for themselves in materials saved within two years of completion.[21]

Books have been filled with anecdotes about the effectiveness of toxics-reduction strategies applied at specific companies. Here are just some of the success stories:

- Riker Laboratories in California is saving $15,000 annually since it replaced an organic solvent with a water-based solvent for coating medicine tablets. In addition, Riker realized a one-time savings of $180,000 by not needing to invest in pollution-control equipment, because the switch to the water-based solvent dramatically reduced air-pollution emissions.[22]
- Chevron USA is now using a high-pressure, closed-loop, hot water cleaning system instead of a system that relied on toxic chemicals. This has resulted in $50,000 annual savings in waste-management costs.[23]
- The Borden Chemical Company installed new equipment to rinse filters and to clean tanks that reduced the discharge of organic solvents into the wastewater stream. This reduced annual reduction in their disposal costs by $48,000.[24]

None of these figures, of course, take into account the very large savings *to society* when toxics are reduced at the point of origin. Society clearly gains from less environmental cleanup, lower medical bills, fewer lost work hours, and reduced property damage. Toxics reduction is a good economic deal for society as well as for industry.

Yet in recent years few firms have systematically reviewed and implemented the available measures. Companies have failed to adopt reduction methods because they lack a full awareness of the alternatives, they focus on short-term profits, they cling to old ways of doing business, and they underestimate the hidden benefits of reduction to the firm. The challenge, as the Congressional Office of Technology Assessment (OTA) stated in a 1986 report, "is to persuade most American waste generators to do what a few companies have already discovered is in their own economic self-interest." The OTA further noted that, with strong encouragement by government to look at source-oriented solutions, firms could achieve a 50 percent reduction in hazardous-waste generation within five years.

Federal Disclosure Measures

Three federal laws require corporations to disclose data potentially relevant to reduction of the use of toxics. First, under the Resource Conservation and Recovery Act, the national law governing hazardous wastes, companies are required to file reports every two years noting what they have done to reduce the production of hazardous waste.

Second, each major polluter's estimates regarding its emissions into all environmental media are submitted annually to the EPA under the TRI. These data are stored in a publicly accessible database system maintained by the National Library of Medicine and are summarized each year in an EPA report.

Third, starting in July 1992, the federal Pollution Prevention Act requires companies to add further information to their TRI filings. Industries must submit to the EPA information on their efforts to recycle and to reduce their use of toxic chemicals. These reports document changes in the patterns of use and waste of every toxic chemical during the previous year and expected changes over the next two calendar years.

None of these laws, however, require any particular toxics-use reduction measures, or, in fact, any toxics-use reduction at all. Instead, they simply establish a minimal level of public disclosure and count on companies to undertake necessary changes voluntarily.

Voluntary Approaches to Pollution Prevention

Aside from mandatory *disclosure* programs under federal law, the principal nationwide initiatives to reduce the use of toxics are voluntary efforts being promoted by the EPA, industry, and citizens groups. Here we note the three programs with the highest profile:

The EPA Industrial Toxics (33/50) Program

In response to a growing public call for toxics reduction, the EPA has decided to promote not additional regulation, but industrial volunteerism. Its Industrial Toxics Program invites companies to reduce emissions of 17 toxic chemicals—by one third by the end of 1992 and by one half by the end of 1995. The program thus has become known as "33/50." Since the reduction of emissions into *all* environmental media is being encouraged, it is implicitly assumed that this program will lead to reduced usage of the 17 chemicals.

The primary incentive for companies to comply is positive publicity. Firms that sign up can proudly assert that they are good "corporate citizens." The EPA helps with the publicity by periodically publishing a list of participating companies and an assessment of their progress. A secondary and less explicit incentive is to avoid regulatory action. Some companies believe that this voluntary program might eliminate the need for the EPA to adopt a stringent multi-media reduction rule for the targeted chemicals.

Self-Policing by the Chemical Industry

According to an opinion poll conducted in March 1990, 75 percent of the public viewed the chemical industry unfavorably; only the tobacco industry had a lower rating. More than 60 percent of the public rated the chemical industry as "very harmful to the environment." In response, the Chemical Manufacturers Association launched the Responsible Care Program, which requires member firms to follow a code of conduct regarding waste reduction and community relations.

One provision of the Responsible Care Code provides as follows:

> Each CMA Member company shall have a pollution prevention program which shall include (among other things):
>
> - Ongoing dialogue with employees and members of the public regarding waste and release information, progress in achieving reductions, and future plans. This dialogue should be at a personal, face-to-face level, where possible, and should emphasize listening to others and discussing their concerns and ideas; and
> - Periodic evaluation of waste management practices associated with operations and equipment at each member facility, taking into account community concerns . . . ."[25]

The CMA's ads even invite neighbors to visit local plants to find out what is going on inside. With its unofficial motto, "Don't Trust Us, Track Us," Responsible Care suggests that chemical companies are now operating with a new spirit of responsibility toward their neighbors. The industry claims to be engaging in self-policing by encouraging firms to clean up their own acts and by using its trade association to put pressure on laggard firms.

The Ceres Principles

Corporate codes of conduct also have been initiated from outside corporate boardrooms. The CERES Principles (formerly known as the Valdez Principles) are a code of conduct advanced by the Coalition for Environmentally Responsible Economics, which includes shareholders, environmentalists, and labor unions. Signators to the CERES Principles pledge to "reduce, and where possible eliminate the use, manufacture or sale of products and services that cause environmental damage or health or safety hazards." Other provisions seek: to eliminate the release of any pollutant that may cause environmental damage to people or the biosphere; to safeguard habitats in rivers, lakes, wetlands, coastal zones, and oceans; and to minimize contributions to global warming, ozone depletion, acid rain, or smog. The CERES Principles also bind signatories to reduce outputs of toxic wastes, to use energy wisely, to market safe products and services, to compensate the public for any damage, and to disclose information about risks they are posing to the public.

All three of these voluntary programs are legally unenforceable, as the CERES Principles explicitly concede: "These principles are not intended to create new legal liabilities, expand existing rights or obligations, waive legal defenses, or otherwise affect the legal position of any signatory company, and are not intended to be used against a signatory in any legal proceeding for any purpose."

A Critical Examination of the Voluntary Programs

These nonregulatory initiatives assume that corporations will take the right action to maximize economic gains and to satisfy public prodding. But is this assumption correct? Or is additional governmental intervention necessary to produce socially desirable results?

The voluntary programs certainly offer some advantages. They require minimal work by regulatory officials. They provide the flexibility for businesses to take only those actions deemed profitable or affordable. They prevent government "meddling" in corporate decisions and operations. They set an agenda, albeit a vague and legally unenforceable one, for businesses to give more attention to reducing both wastes and the use of toxics. And they at least raise the issue of accountability by acknowledging the need for increased participation and democratization in corporate decisionmaking. The theme of increasing external accountability by disclosure and "dialogue" is raised in Responsible Care and in the CERES Principles, and is an acknowledged need in the 33/50 Program. Together, these programs open a window of opportunity for workers and the public to increase corporate responsibility.

Another positive aspect of the voluntary strategies is that they engage the parties "closest to home" in cooperative efforts to solve a community's environmental problems. Since the federal government cannot possibly address all the problems on behalf of the 30,000 communities in America, this seems to be a needed and often missing element of federal environmental strategies.[26]

Yet, lacking the foundation of legal rules, these voluntary strategies run up against the forces of the marketplace. As long as corporations can secure quick profits elsewhere, they will resist making investments in the long-term sustainability of a local manufacturing facility, despite kinds words,

A Typical Experience with "Responsible Care"

Our citizens group, which we call "Good Neighbors," grew out of an incident that occurred on 25 June 1991 at the Rhone-Poulenc plant in Sedalia, Missouri. A batch of monomer, diallyl maleate, went bad. It "polymerized" and ruptured the relief valve, then blew up, spewing the contents directly into the atmosphere. An evacuation area was established by the local Fire Chief. This incident included explosions of fireballs about three-and-a-half hours after the initial release occurred and a fire, which occurred two hours later. That occurred twelve months and one week after a similar incident involving another monomer. The 1990 incident involved an evacuation of over 1500 people and lasted longer. It emitted steadily for around twelve hours, then off and on for two more days. The evacuation was in place only for the first six hours.

After the 1990 incident the company held one public meeting to answer questions, established a hot line for calls, paid for carpet and furniture cleaning, and [paid] medical bills and lost wages for evacuees. There were several civil suits filed. After the 1991 incident the company did nothing except make a few comments to the press.

Our group formed one week to the day after the 1991 incident and we began having weekly meetings. We began communicating with the company through letters almost immediately. We gathered all the information available locally concerning the incident through the Local Emergency Planning Committee (LEPC) office. We received a great deal of press coverage and our meetings began being attended by employees of Rhone-Poulenc, including engineers and other management personnel. Theirs was an adversarial role. The company referred to us as "hysterical neighbors."

We contacted CMA concerning the Responsible Care program and they sent a packet that contained all the self-audit forms sent to companies. We asked to see Rhone-Poulenc's, since they said they belonged to the Responsible Care program. They said this was not available. We asked to see their worst-case accident scenario and they said they did not have one. Most of their responses to our queries were very brief and incomplete, and in almost moral commitments, or pledges to be good neighbors. Volunteerism is not proving to be strong enough to break the bad habits of many firms that have not even studied options to prevent pollution. In fact, volunteerism may only be postponing the beginning of better solutions.

Toxics Reductions Are Limited

How are the voluntary programs stacking up? The EPA's public relations suggest that a major breakthrough has occurred. It called the 33/50 Program a success in February 1992 because companies had voluntarily pledged to reduce emissions by 304 million pounds, which, according to

every case, did not answer the questions asked. In fact they wrote the LEPC office saying that any further inquiries from our group should be directed to Missouri Department of Natural Resources, not Rhone-Poulenc.

Next, they attempted to include us in the formation of a Community Forum. Since I have always acted as spokesperson for our group, I was contacted by a company consultant and asked if I would serve on the Community Forum. After consulting with several environmental groups I agreed to give it a try. I was the only female on the forum, and the only person not in a position of power. The others were a physician, a minister, and a mayor — none of whom lived anywhere near the affected area. The members were selected by the plant manager.

The only meeting I attended began with the plant manager saying that Rhone-Poulenc needed to help the community learn to trust them and that I was on the forum because if they could get me to trust them, then they were sure they could get the other members of the community to trust them. It was obvious that I was the object of the forum discussions. During the two-week period that Rhone-Poulenc thought of me as being on the forum, I received daily telephone calls from the plant manager about individuals in our group who were making comments he did not like. It was obvious to me that I was being used as "their girl." I withdrew from participation when the facilitator phoned me to say the plant manager had asked her to call me, because he was concerned about a press conference that had occurred in my front yard. Since that time we have continued to correspond with Rhone-Poulenc. We asked to be able to go to the facility and we wanted to bring our own "experts." They said there was nothing in the federal Right to Know Act that made this mandatory and they refused.

We have yet to receive adequate information about what substances the community has been exposed to. Because the substances released in the incidents were specialty chemicals manufactured only by this facility and about which no health information is obtainable, they are called non-hazardous. They do not appear on the SARA Title III lists, even though their ingredients all do.

—Kathy Grandfield

the EPA, is the equivalent of "10,000 fully-loaded 52 foot tanker trucks, stretched end to end for 100 miles."

But to put this "success" in another light, compare it with the OTA's projection that a 50 percent reduction in all waste produced nationally over five years would be possible with a strong government effort. The voluntary initiative of 33/50 so far has yielded only commitments to reduce emissions 14 percent from 1988 to 1995. The difference between a 50 percent reduction in five years and a 14 percent reduction in eight years highlights the gap between a government which talks softly and carries a big stick and an EPA which advertises loudly and carries no stick at all.

The companies that have comprehensive toxics-use reduction programs are barely visible on the map. Even though the government provides support for voluntary waste reduction in the form of information or capital, the number of companies that are systematically and aggressively implementing toxics-use reduction is limited. One of the oldest programs for voluntary waste reduction has been in North Carolina, where the state government has provided technical and financial assistance to interested companies. Even there the results have been discouraging. Only eight percent of industries in North Carolina were found to have any waste reduction programs in place.[27]

Unaccountable Corporate Performance

Toxics-use reduction measures may lower the risks to the public and to workers and may improve the sustainability of some local industries. But in the absence of public scrutiny there is no assurance that corporations will proceed as quickly as is economically possible to reduce their use of toxics. Moreover, recent case studies indicate that toxics-use reduction may be used as an excuse to cut jobs or to increase demands on the land and the stress levels of workers. Environment-friendly measures are not necessarily worker-friendly.

Demands for pollution prevention are themselves sometimes turned against plant workers in a new form of job blackmail. The firm's managers give the employees an ultimatum: "Find a way to end the pollution or we will end your jobs!" At the Robbins Company in Massachusetts, management told workers that unless the firm achieved *zero* waste production in electroplating, their jobs would be cut and the plating work "contracted out" to another firm. Fortunately, the workers were able to identify ways of cutting waste. But in many other settings this approach would mean "blaming the victims" for management's faulty investment decisions of the past. Additional capital expenditures or investments in research and development may be needed, all of which are usually out of the control of workers.

Environmentally beneficial innovation does not always mean more jobs or higher quality employment. The traditional strategies employed by industry to modernize and increase plant efficiency involved doing more work with fewer workers and resources. There is some evidence, however, that the expanded use of toxics has replaced more jobs than it has created. For example, many chemicals were introduced to reduce the amount of manual labor needed in production. The proliferation of pesticides, fertilizers, and machinery in agriculture dramatically reduced the number of farmers needed.[28] As an industry moves away from toxics, there is a possibility that the amount of labor needed will increase. But will management require existing workers to do the new jobs in addition to their previous ones, without adding workers to meet the added labor needs?

None of the voluntary initiatives adequately address the needs of communities or the workforce. Members of the public and workers should have the right to the kinds of information and oversight that can ensure progress and prevent abuses.

The biggest part of Responsible Care seems geared to rehabilitating the industry's low public image through a $10 million advertising campaign. Despite the rhetoric of accountability, the program is not yet improving industry performance. The United States Public Interest Research Group (USPIRG) and member PIRGs around the United States decided to test the willingness of CMA members to be accountable.[29] They contacted 192 CMA member facilities in 28 states and asked nine basic questions relating to chemical usage, emissions, and reduction. Only 19 of the facilities (17 percent) answered each of the nine questions. At 42 percent of the facilities, PIRG interviewers could not reach anyone to answer their repeated calls. At 27 percent of the facilities, the company contact either could not or would not answer any of the questions. At 55 percent of the facilities the contact could only answer some of the questions, and 29 of these answered less than half of the questions. At a 3M facility in Minnesota, a caller was told that the facility does not give information out to citizens. When the caller asked, "How would citizens go about getting information about your facility?" he was told there was "no good way."

As part of their self-policing effort, CMA firms rate their own behavior. Eventually the trade association itself can serve as an enforcement body by expelling any member that fails to live up to the requirements of the CMA Code. But one official of the CMA who investigated citizen complaints about a member company noted that it will be at least three years before the CMA even considers expulsion or discipline of member firms. Even though the CMA's Public Advisory Panel recently reached a consensus that "a system of independent evaluation of industry performance is critical to making the program work and ensuring credibility," the likelihood of stringent self-regulation within the trade association is dubious—like the proverbial fox guarding the chicken coop.

Lacking a genuine oversight process, CMA members are continuing to block genuine public accountability, sometimes even invoking Responsible Care as they do so. The CMA's advertising announces proudly that companies are now inviting concerned citizens to tour their plants and are establishing Community Advisory Panels (CAPs). But Monsanto officials in Texas told a citizens' group that the company's invitation to tour their Superfund site applied to neighbors only and did not include the group's hand-picked experts. The group decided to decline the invitation. Another CMA member, GenCorp, which runs a latex polymers plant in Mogadore, Ohio, established a fifteen-member CAP. While one of the panel's members raised genuine health concerns, the others allowed the financial interests of the company to outweigh any health effects from the company's processes. The lone activist eventually quit the panel in disgust.

"Locking in" Nonsolutions

Historically corporations have advanced innumerable "least cost" environmental solutions which themselves have created new environmental problems. For instance, commercial hazardous-waste landfills proliferated when corporations stopped dumping chemical wastes on their own property or directly into local receiving waters. As the hazardous-waste landfills of the 1970s have become the Superfund sites of the 1990s, an increasing portion of the industrial waste stream is being burned in incinerators, industrial boilers, and cement kilns. Environmental groups now oppose these solutions, which disperse the toxic waste stream into the air through smokestack emissions and onto the land in the form of residual ash. The voluntary programs, since they do not necessarily prevent all kinds of pollution, will themselves produce a new generation of environmental problems.

For instance, two out of three voluntary programs focus on "*waste* reduction" rather than "toxics-use reduction." Some federal, state, and corporate officials consider "waste reduction" to include off-site recycling or even incineration. Yet the burning, disposal, treatment, or out-of-process recycling of wastes can all create very substantial exposures and risks to local communities. Overall toxics reduction is clearly the preferable goal, because it reduces both the use of toxic chemicals and the production of toxic wastes at the point of origin.

The CMA's own newsletter opposes toxics-use reduction as an idea of "extremists,"[30] perhaps because the concept threatens to curtail the use of the very products that the industry wants to produce and sell. Waste reduction as advanced by the CMA, the 33/50 Program, and the CERES

Principles do not prevent workers and consumers from being exposed to toxics. The voluntary programs seem to be geared toward heading off mandatory toxics-use reduction. While the CMA wears a "responsible" face in public, its politicking and public relations efforts show another face entirely. CMA member companies have lobbied Congress not to enact a Right to Know More Bill, which would grant citizens access to information about more facilities and would require companies to conduct toxics-use reduction planning. Some company lobbyists have even gone so far as to assert that voluntary programs, such as Responsible Care, make any new legislation unnecessary. Meanwhile a public relations plan prepared for the Clorox Corporation (a member of CMA) calls for attacking Greenpeace as a group of violent and self-seeking "eco-terrorists," suing newspapers that advocate the use of nontoxic bleaches, and recruiting scientists to question studies that link chlorine to cancer.

In contrast to the commitment to 14 percent waste reduction reached under 33/50, a growing chorus of policy analysts, such as those from the U.S.-Canada International Joint Commission, are joining with environmental groups like the National Toxics Campaign Fund and Greenpeace to demand that the EPA completely ban many of the 17 chemicals targeted by the industrial toxics program. Examples of persistant 33/50 substances that environmentalists seek to ban because of their environmental persistence include lead, trichloroethylene, trichloromethane, tetrachloroethylene, and dichloromethane. But the EPA's 33/50 effort implicitly commits the agency not to ban these chemicals for five years.

Why Voluntary Programs Cannot Succeed

Voluntary programs not only are failing now but also are doomed to fail in the future. There are three reasons why:

The Constraint of Short-Term Profit

The single largest impediment to enhanced environmental performance by corporations is their obsession with short-term profit. This is the conclusion reached by corporate CEOs and environmental managers themselves. Tufts University's Center on Environmental Management[31] recently surveyed the CEOs and directors of the environment, health, and safety divisions of 98 companies. When asked what is preventing their companies from doing a better job, the highest number of respondents, 53 percent, cited the emphasis on short-term profitability.[32] This is reinforced by such day-to-day factors as promotional decisions and fluctuating stock market prices, which tend to be affected by recent profits and losses that are measured monthly or quarterly.

Because of concerns about short-term profitability, firms are reluctant to invest in research or production changes where the payback may be long term. Unfortunately, not all improvements in toxics-use reduction can be squeezed inside the narrow window of short-term paybacks. Sometimes increased efficiency requires a longer time frame to pay back. Other times the payback is to society or to the workforce, not to the company's profits. Stopping pollution, creating more long-lasting jobs, and moving toward sustainable manufacturing may yield no new profits to a firm, yet they may be of great economic benefit to the workforce and to the community.

An integrated approach to pollution prevention means that firms must do more than add-on equipment. It often requires restructuring the way products are made and jobs are performed. It may demand a "stay put" mentality—a commitment to keep the firm operating in the community—that goes against the recent onslaught of plant closings and migrations of manufacturing facilities to Mexico and other Third World locations in the name of free trade.

Environmental laws have yet to constrain a firm's inclination to choose short-term profits over longer-term sustainability. While some environmental laws have stated "zero pollution discharge" as a long-term goal, the actual legal requirements do not go so far. Even though adding *no* pollution to the environment would be the wisest course for many pollutants, the rules now dictate how much pollution and what kind of dumping methods are acceptable. In an economy that is competing for short-term profitability, there is strong pressure on managers to find the cheapest way to produce within those dictates. Production facilities have been engineered to meet the law—and not to produce the least pollution.

If the goal of zero discharge were effectively written into law, firms would be driven to address the underlying causes of their pollution. The amounts and types of polluting substances that firms use and the practices that create waste would become the principal foci of their environmental decisionmaking. By tolerating some pollution, the law encourages industries to continue dumping waste and not to press forward to find the cleanest, waste-free production processes. Wasteful production also has adverse economic repercussions. Pollution at Superfund sites has strapped the United States with a cleanup bill of several hundred billion dollars. And toxic pollution adds to the national health-care crisis by spreading illness and raising medical bills.

Other Institutional Impediments

Unlike pollution control add-ons, toxics-use reduction requires firms to reevaluate their entire way of making products. They must review and evaluate the processes in which toxics are used and the range of alternatives for potential reduction. Most companies are not yet gathering this needed information. Indeed, reports by the OTA and the EPA have found serious psychological and institutional impediments to a system-wide reduction of toxics and waste. For example:

- Businesses fail to take into account avoided costs in calculating relative rates of return.
- Concerned about maintaining product quality, industries are reluctant to tamper with production processes and rarely consider how to make the product without toxic chemicals.
- Environmental and safety engineers are trained and directed to design solutions at the "end of the pipe," or technological fixes, while the engineers with the production-line information are directed not to address safety concerns at all.

To create a holistic approach sufficient to identify and implement reduction options, the various employees charged with production, profitability, and safety must bring their thinking together. This requires a big change in old habits, and the psychological hurdles to overcome are substantial.

For example, a company environmental engineer designed a process ten years ago using a control strategy.[33] A closer look would indicate that using pollution control instead of redesigning the entire plant wastes fifty thousand dollars of materials per year. But will the company take that closer look? Not if the engineer can avoid it. He doesn't want to answer embarrassing questions about why he did not curtail waste generation ten years ago. He may even fear that his job is on the line. Now that the firm has sunk its investment into pollution control equipment, pollution prevention might be too expensive.

Encouraging prevention only gets harder, and the organizational resistance more severe, as the corporation moves from simple actions to more complex matters. Changing production processes and putting greater resources into research and development may take even more commitment and public accountability.

Voluntary Initiatives Cannot End Noncompliance

Innumerable U.S. corporations treat a potential penalty for an environ mental violation as an ordinary cost of doing business. The adoption of a code of conduct by an entire polluting industry or the adoption of an "environmental policy" by a recalcitrant lawbreaker will not suffice to end criminality and prolonged noncompliance where such behavior "pays."

Thus, even if pollution prevention is cheaper than full compliance with the standards, the cost-benefit balance will tip against toxics-reduction measures if the firm is not really required to comply with the standards.

At least one empirical study has indicated that codes of conduct do not curtail illegal behavior. A study of illegal acts committed by companies from 1973 to 1980 found no relative decrease in violations by the companies that adopted such codes. In fact, the study, which examined environmental violations as well as labor, financial, and trade violations, found that "the presence of specific categories of code content appears to be associated with a higher number of a corporation's violations, rather than the opposite. The type of industry and size (large) of the corporation have a far greater impact on violations than does a code of ethics."[34] One reason why codes have had so little impact is that they have traditionally been unenforceable; neither the government nor the public can ensure compliance.

A Strategic Organizing Response

How can grassroots organizers respond to the growing toxics crisis? A first step is to build upon the efforts of grassroots groups during the last decade to make corporations accountable to local communities around environmental issues. In this emerging model, the community and the workforce would exert genuine oversight to ensure that measures for toxics-use reduction are advanced expeditiously and in a manner that supports the needs of the community and the workforce. Across the United States neighborhood groups have mounted successful campaigns to get local industries to allow citizen inspections and to negotiate with the companies directly for enhanced environmental-protection measures.

In an early example of the application of this strategy, the National Toxics Campaign Fund worked with the Quinsigamond Village Health Awareness Group in Worcester, Massachusetts, to wage an extensive campaign to clean up the Lewcott Company. Neighbors suspected that odorous emissions were causing a variety of local health problems, including a high incidence of respiratory illnesses and cancer. The plant also posed substantial fire hazards, because of the storage of large quantities of flammable chemicals so close to people's homes. After pickets, bad publicity, and challenges to the company's licenses, Lewcott officials decided to negotiate. A historic accord was reached, in which the company agreed to move its dangerous operations to another facility a few miles away. The agreement also granted the community group, in the interim, the right to inspect the company and to assess penalties of $5,000 per day for odors emitted in the future. In return, the group committed itself, among other things, to remove the anti-Lewcott signs that had been posted on their front lawns along the busy street where the plant was located. The agreement was enforceable through binding arbitration and was ratified by residents of the community in a neighborhood-wide meeting.

Besides empowering local citizens to solve local environmental problems, this "Good Neighbor Strategy" has often resulted in improved safety and plant modernization, especially when it has also involved plant workers—"win-win-win" solutions in which neighbors, corporations, and workers all benefit. This could become a new paradigm for problem-solving partnerships among local citizens, workers, and corporations in which firms commit themselves to address toxics-use reduction, to grant access to citizens and their experts to assess cleanup progress, and to meet other locally defined needs.

An example of where workers were integrally involved in toxic clean-up occurred at Sheldahl Inc., a manufacturer of electronic circuitry and laminated products that uses methylene chloride (MeCl) to cement copper circuits to a base material. The company released methylene-chloride emissions near the town of Northfield, Minnesota, for over 25 years. In 1987 Sheldahl ranked as the nation's 45th largest industrial emitter of eleven known or probable carcinogens in the TRI. About 600 tons of methylene-chlorided was released into the air that year, and this was merely a fraction of the actual amounts to which workers who handled the chemical were exposed. When the federal government released this information, it caused an uproar. It was the first time the federal government told citizens who was polluting their community. The TRI data encouraged citizens to act. Two citizens' groups, the Northfield Air Toxics Study Group (ATSG) and Clean Air in Northfield (CAN), were established. A series of meetings involving these groups, the company, and the Amalgamated Clothing and Textile Workers Union (ACTWU) led to a unique negotiation process that may set a precedent for other plants in the future.

The ACTWU feared that citizens' concerns for air quality might lead to the plant being closed. So on November 1, 1989, when the ACTWU contract expired, the union negotiated a new collective bargaining agreement and a written commitment from Sheldahl to eliminate 90 percent of methylene-chloride emissions by 1993, to reduce usage of the chemical by 64 percent by 1992, to hold monthly meetings with community groups and the union to assess progress, and to prioritize capital expenditures to eliminate the chemical altogether and to find a non-toxic alternative.

Thanks to the combined effort of the citizens' groups and the ACTWU, the state required Sheldahl to reduce cancer risks further and to lower methylene-chloride emissions by 93 percent by 1995. The Pollution Control Agency stated that there would not have been any restrictions in Sheldahl's permit if the union and the citizens' groups had not intervened. Sheldahl had previously claimed that reductions were not possible within the desired time frame.

Together, the examples of Lewcott and Sheldahl demonstrate a model that goes beyond voluntary toxics-use reduction in which local concerned citizens and workers would have rights: (1) to inspect facilities with their own experts; (2) to require the firm to assess possibilities for toxics-use reduction; (3) to have access to all information developed by the firm regarding toxics-use reduction; and (4) to oversee implementation of all the options for toxics-use reduction that prove feasible.

Alliances of labor, community, environmental, and church organizations may be able to achieve more successes together than any of these groups could achieve on their own. The agenda of such a coalition is likely to extend beyond reducing the use of toxic chemicals and toward creating "sustainable industry" for the community. It would include other objectives such as:

- seeking financial disclosures by the company to allow a realistic appraisal of the economic feasibility of various toxics-use reduction measures;
- preventing less experienced, nonunion workers from undertaking certain jobs that pose accident risks to the entire community;[35]
- assessing whether a local industry is investing enough in its local plant and workforce to ensure that the plant will remain viable and continue providing jobs for the community; and
- conducting periodic citizen/worker inspections to identify occupational and environmental hazards.

Public Policy Recommendations

While alliances among workers, communities, and citizen groups are valuable, there is much that the government can do at the national, state, and local levels.

Expand Public Rights

Citizen/worker negotiations with local manufacturers will be handicapped until all members of the public can hold corporations accountable as a matter of right. This means expanding the kinds of information to which the public is legally entitled, guaranteeing the public access to facilities for inspections, and establishing public oversight committees with the power to force manufacturers to reduce their reliance on toxics and to make plants more sustainable. Some of the groundwork has already been laid. The Hazard Elimination through Local Participation (HELP) Act, which was proposed in New Jersey in 1989, would have given neighbors and workers a right to inspect facilities. Legislation enacted in several states and proposed in Congress would require industries to prepare and disclose plans for reducing the use of toxic chemicals. Rights also can be advanced by other means. For instance, legislation or court orders could make corporate charters and bylaws, environmental permits, and state-government approvals conditional on giving the public greater rights.

Require Firms to Study Toxics-Use Reduction

A few states such as Massachusetts and Oregon have required industries to establish programs for toxics-use reduction. For example, the Massachusetts Toxics Use Reduction Act requires each industry that uses more than a threshold quantity of certain toxics to prepare a use-reduction plan which will: (1) quantify each hazardous substance used, generated, released, or transferred off-site during the year; (2) identify all available methods for toxics-use reduction for each substance; (3) analyze the costs, savings, and overall feasibility of each identified method; (4) list the methods selected; and (5) establish a timetable for implementing these methods.[36] Other states, and the federal government, should enact similar laws.

Establish Structured Participatory Decisionmaking Processes

Getting neighbors, workers, and plant managers to sit around the same table to discuss toxics-use reduction poses a serious challenge. Plant managers tend to resist real power-sharing with workers and with the local community. Consequently, some form of committee structure which addresses toxics-use reduction and the sustainability of the manufacturing process should be established for each firm as a matter of law. Several proposed pieces of federal and state legislation (such as the federal OSHA Reform Act) would establish oversight committees involving management, workers, and neighbors and would protect nonunion workers against harassment for participating in such committees.

Enact Strong Land Use Controls to Buffer Hazardous Operations

Federal, state, and local laws should prohibit the location of hazardous operations in close proximity to people's homes. A buffer-zone ordinance proposed in San Diego, for example, would preclude the placement of large users of toxic chemicals within 300 feet of residential areas.

Sunsetting Toxic Chemicals

The most effective means to eliminate the harmful effects of toxic chemicals is to prohibit their production and use altogether. A growing number of policy analysts and environmental groups are advocating bans or phaseouts of various substances. Highest on the target lists are chlorinated solvents such as trichloroethylene, trichloroethane, tetrachloroethylene, and dichloromethane, all of which tend to persist and accumulate in the food chain.

Thus far, few harmful chemicals have actually been outlawed, and the drive to ban these chemicals received a major setback when the Fifth Circuit of the U.S. Court of Appeals struck down the EPA's attempted ban on asbestos in October 1991.[37] The court held that the EPA had not adequately considered the availability of alternative regulatory measures short of a ban, nor the environmental effects of alternatives to asbestos. The Toxic Substance Control Act (TSCA), under which the ban was proposed, sets forth a rational framework to consider a wide range of regulatory measures. But the court's holding creates formidable obstacles for the EPA to overcome if it attempts to ban chemicals in the future.

Those who want to ban chemical products will need to consider vehicles other than TSCA. Perhaps special federal legislation can be passed that bans or systematically phases out a single chemical or a whole group of compounds. In the absence of a clear public consensus in favor of a ban or phaseout, there still may be public support for measures to discourage the use of a toxic chemical, like an escalating tax. The Clean Air Act established just such a tax on the production of ozone-destroying chemicals. Taxes have the advantage of providing a funding pool that can provide compensation to workers displaced by the phase-outs.

Capital Punishment for Corporate Crimes

Perpetual lawbreaking by some corporations is not just a health and environmental problem, it is also an economic problem. The lack of compliance by some firms discourages toxics-use reduction and creates serious competitive disadvantages for law-abiding firms. To increase compliance, new enforcement strategies are needed. Where corporations engage in wanton behavior towards workers or environment, new legal remedies should be used to treat the corporation itself as a criminal. Courts should be encouraged, through litigation and legislation, to revoke the charters of misbehaving corporations. This corporate "death penalty" should be reserved, of course, for the worst of corporate criminals. Upon revoking the charter, a court should devise the best possible strategy to preserve jobs and workers' income. Where equipment and plants are salvageable, the court should remove existing managers and sell the corporation's assets to the bidder who proposes the best plan to protect the interests of workers, previous shareholders, the community, and the environment. Where equipment and plants are not salvageable, the court should liquidate whatever assets remain and distribute them fairly among shareholders, the workforce, and the community.[38]

Other more modest ways to improve enforcement include replacing industrial "self-monitoring" with monitoring by environmental auditors who are accountable to neighbors and workers, eliminating impediments to citizen suits, altering environmental statutes to make it easier to assess civil and criminal penalties directly against top corporate officers, and ensuring that civil penalties assessed against corporations recapture a multiple of the profits (treble damages or more) reaped through noncompliance.

Eliminate Liability Shields for Corporate Executives and Owners

The "voluntary" economic incentives for toxics-use reduction would be stronger were the individual assets of economic and management decision-makers on the line. Corporate and liability law should be reformed to make it easier to hold both major shareholders and corporate managers personally liable. This will encourage more responsible behavior by these parties.

Expand Publicly Available Technical Resources for Toxics-Use Reduction

The ability of investors, consumers, workers, and neighbors to induce sustainable corporate behavior will be hampered as long as they are unable to obtain adequate information and technical resources. The government should provide or fund these accountability tools. Federal and state agencies might provide technical assistance grants, training programs, or technical personnel to community and workplace activists, or they might expand the availability of information about corporations and their products. Better information would help equip the public to negotiate more effectively with corporate managers. It also would allow investors and consumers to "vote with their pocketbooks" for the most environmentally sensitive corporate stock and product offerings.

Promote Federal Intervention for Innovation

The government must encourage private sector innovation toward more environmentally sound technologies. While community-level pressure may move some individual corporations to scrutinize their behavior and to invest more in research and development, only the federal government can mobilize resources on a large enough scale to advance corporate innovation for toxics-use reduction. For example, in 1992 Congress proposed earmarking a portion of its funding for Sematech, the public-private consortium for semiconductor research, for environmentally sound production research. The bill would have allocated $10 million of the annual $100 million grant to develop environmentally sound alternatives for production. The federal government also is one of the largest single buyers in the marketplace. It can create a predictable market for many environmentally sound products. One priority area which has already received some public policy attention is for the military to reduce its purchase of goods produced with ozone-destroying compounds.

Reform Economic Development Programs

Local, state, and federal economic development programs that provide industries with capital assistance should condition their loans, grants, or loan guarantees on toxics-use reduction. Unlike past subsidies, government agencies must demand specific and measurable commitments (a company might promise to create 100 jobs for local residents over a ten year period without using toxic substances). If such commitments are breached, the recipient should be obliged to promptly pay back, with interest, the entire subsidy received.

Ensure Fair Economic Transitions

Not every plant can be saved. In fact, from an environmental perspective there are many plants that are beyond hope and *should* be shut down. For outdated or specialized pieces of equipment that are only suited to producing or using banned or phased-out chemicals, the only desirable "conversion" may be to recycle them as scrap metal. The workers in these plants, however, must not be thrown onto a scrap heap. National legislation to address worker displacement must be an integral part of toxics-use reduction. The Oil, Chemical and Atomic Workers International Union (OCAW), which represents workers in the most toxic industries in America, has called for a "Superfund for Workers." If a plant is closed for environmental reasons, the fund would support workers at existing wage levels and pay for up to four years of education to ensure their smooth transition to comparable paying jobs.

Reduce Global Whipsawing through International Standards

International free trade regimes like the North American Free Trade Agreement (NAFTA) enable corporations to play one community off against another to achieve the least restrictive environmental and labor conditions. Organized citizen pressure and public education can help to establish local structures which make this corporate strategy harder to employ. But until more uniform standards are enacted that apply everywhere, whipsawing will continue within and outside the United States. National and international standards that would require all industries to study and reduce their use of toxics remain a fundamental goal. The absence of such standards to protect the environment and workforce in other countries means continuing pressure on U.S. industries to cut corners. The 1992 United Nations Conference on Environment and Development was a missed opportunity to produce international environmental standards for trade. As the worldwide grassroots movement for toxics-use reduction and sustainable industry continues to grow, this issue will move to center stage.

As this essay goes to press, the Clinton Administration prepares to take office with a promise of new ideas and a new generation assuming the mantle of power. In his election campaign, Bill Clinton emphasized the need for *both* investment in America and empowerment of local communities. Toxics policies represent one important area wherein new ideas can be applied immediately. In lieu of the trend toward "volunteerism," the new Administration could put into effect a combination of investment, enforcement, trade, accountability, and empowerment policies that create truly sustainable industries for the United States — industries that are clean, stable and fair to their local stakeholders. Further regression toward more volunteerism would only delay genuine protection from toxics of the environment.

Notes

1. *Changing Course, A Global Business Perspective on Development and the Environment*, (Cambridge, MA: MIT Press, 1992), pp. 2-3.
2. See S. Lewis and M. Kaltofen, *From Poison to Prevention* (Boston:National Toxics Campaign Fund, 1987).
3. Samuel S. Epstein and Ralph W. Moss, Letter to the Editor, *New York Times*, 16 June 1991.
4. Keith Schneider, "Toxic Pollution Shows Drop in '89," *New York Times*, 17 May 1991, p. A32.
5. *Ibid.* See Joel Hirshorn and Kirsten U. Oldenburg, *Prosperity Without Pollution*, (NY: Van Nostrand Reinhold, 1991), p. 88. The Community Right to Know Act required industries to disclose the amount of toxic chemicals discharged to air, water, and land. In 1989, according to the EPA, 22,650 industrial sites across the United States reported releasing 5.78 billion pounds of toxic chemicals into

the environment. Total emissions are actually higher, since the approximately 330 hazardous substances monitored by the EPA's Toxics Release Inventory do not include more than 500 toxic chemicals regulated under other environmental laws. In addition, many companies flout the Right to Know Act and have not yet filed the legally required information.

6. The U.S. government also engages in toxic dumping on military bases, for example. It is estimated that there exist approximately 14,000 uncontrolled military waste sites. See Lenny Siegel, *The Military Toxics Legacy* (Boston: National Toxics Campaign Fund, 1990).
7. Peter Montague, "Earthly Necessities: A New Environmentalism for the 1990's," *The Workbook*, 16: 2 (Summer 1991), p. 58.
8. *Ibid.*, p. 59.
9. See National Research Council, *Environmental Epidemiology: Public Health and Hazardous Wastes* (Washington, DC: National Academy Press, 1991).
10. William J. Blot and Joseph F. Fraumeni, Jr., "Geographic Patterns of Lung Cancer: Industrial Correlations," *American Journal of Epidemiology*, 103 (1976), pp. 539-550, cited in *Hazardous Waste News*, 274 (26 February 1992).
11. S. Strawn and M. Legator, "Epidemiology and Toxic Torts: Animal Studies Yield Valid Insights," *Trial* (April 1991), p. 61.
12. Bengt B. Arnetz, "Mortality among Petrochemical Science and Engineering Employees," *Archives of Environmental Health*, 46 (July/August, 1991), pp. 237-248, cited in *Hazardous Waste News*, 273 (19 February 1992).
13. Epstein and Moss, Letter to the Editor.
14. For example, since 1950, the incidence of cancer per 100,000 U.S. citizens has risen by 42.2 percent. U.S. National Cancer Institute, *Cancer Statistics Review*, NIH Pub. #90-2789. Between 1980 and 1987, the prevalence of asthma increased 29 percent among Americans. Montague, "Earthly Necessities," p. 60.
15. International Joint Commission, *Sixth Biennial Report on Great Lakes Water Quality*, 1992.
16. AFL-CIO, *Death on the Job: The Toll of Neglect* (Washington, DC, AFL-CIO: May 1992).
17. National Toxics Campaign Fund, "Destroying Our Nation's Defenses: A Citizen Indictment of the Environmental Protection Agency and the Reagan-Bush Administration" (Boston: NTCF,1988), p. 32.
18. "Destroying Our Nation's Defenses," pp. 32, 35.
19. Charles Lee, *Toxic Waste and Race* (NY: United Church of Christ, 1987).
20. Bruce Ackerman and Richard Stewart, "Reforming Environmental Law" *Stanford Law Review*, 37, (1985), p. 1333.
21. Bob Pojasek, interview with author.
22. Institute For Local Self-Reliance, *Proven Profits from Pollution Prevention* (Washington, DC: ILSR, 1986), p. 82.
23. *Ibid.*, pp. 221-222.
24. Duey Sarokin, *Cutting Chemical Wastes* (NY: Inform, 1987), pp. 187-192.
25. Chemical Manufacturers Association, *Pollution Prevention Code of Management Practices* (Washington, DC: Chemical Manufacturers Association, September 1991).

26. For a community-based approach, see "The Communitarian Platform," *The Responsive Community*, 2:1 (1991-1992), p. 4.
27. North Carolina Environmental Defense Fund "Drawn with the Wind: Toxic Air Emissions Across North Carolina," (Washington, DC: Environmental Defense Fund, 1989), p. 17.
28. Barry Commoner, *Making Peace with the Planet* (NY: Pantheon, 1990).
29. Carolyn Hartman, *Trust Us, Don't Track Us*. (Washington, DC: USPIRG 1992).
30. "Extremists Urge Curbs on Toxics," *CMA News* (July/August 1992), p. 4.
31. Ann Rappaport and Margaret Flaherty, *Multinational Corporations and the Environment: A Survey of Global Practices* (Medford, MA: Tufts University Center for Environmental Management, 1991).
32. Management structure was cited second, with 42 percent of the respondents identifying this problem. Limits of technology was cited least frequently, by only 20 percent.
33. Example derived from Hirshorn and Oldenburg, *Prosperity Without Pollution*,
34. *Ibid*.
35. M. Clinard and P. Yeager, *Corporate Crime*, cited in H. Pitt and K. Groskaufmanis, "Minimizing Corporate Civil and Criminal Liability: A Second Look at Corporate Codes of Conduct," *Georgetown Law Journal*, 78 (1990), p. 1559.
36. Massachusetts General Laws c. 21F.
37. *Corrosion Proof Fittings, et al., petitioners v. the Environmental Protection Agency and William K. Reilly, Administrator, respondents*, case no. 89-4596, U.S. Court of Appeals for the 5th Circuit, 947 F 2d. 1201; (18 October 1991.)
38. Idea based on author's discussions with Richard Grossman; see also Ian Ayres and John Braithwaite, "Designing Responsive Regulatory Institutions," *The Responsive Community*, 2:3 (Summer 1992), p. 41.

Appendix A

The Triple Revolution

A Letter

The following letter was sent on March 22 by The Ad Hoc Committee on the Triple Revolution to President Lyndon B. Johnson. The White House reply from Mr. Lee White, Assistant Special Counsel to the President, was received shortly. The letter to the President, together with the Report, was also sent to the Majority and Minority leaders of the Senate and the House of Representatives and to the Secretary of Labor. Text follows:

March 22, 1964

Dear Mr. President:

We enclose a memorandum, The Triple Revolution, for your consideration. This memorandum was prepared out of a feeling of foreboding about the nation's future. The men and women whose names are signed to it think that neither Americans nor their leaders are aware of the magnitude and acceleration of the changes going on around them. These changes, economic, military, and social, comprise The Triple Revolution. We believe that these changes will compel, in the very near future and whether we like it or not, public measures that move radically beyond any steps now proposed or contemplated.

We commend the spirit prompting the War on Poverty recently announced, and the new commissions on economic dislocation and automation. With deference, this memorandum sets forth the historical and technological reasons why such tactics seem bound to fall short. Radically new circumstances demand radically new strategies.

If policies such as those suggested in The Triple Revolution are not adopted we believe that the nation will be thrown into unprecedented economic and social disorder. Our statement is aimed at showing why drastic changes in our economic organization are occurring, their relation to the growing movement for full rights for Negroes, and the minimal public and private measures that appear to us to be required.

A complete list of the individuals who signed this letter can be found on page 159.

THE WHITE HOUSE
WASHINGTON

April 6, 1964

Dear Mr. Ferry:

The President has asked me to thank you for your letter of March 19, in which you enclose the memorandum, The Triple Revolution, drawn up by your Committee.

In recent months the President has taken a number of steps addressed to the problems discussed in your memorandum — poverty, unemployment, and technological change. He has committed this Administration to an unrelenting war on poverty and, as you are of course aware, has submitted to the Congress major new legislation requesting the necessary weapons for the prosecution of this war. On December 21 he established the Committee on Economic Impact of Defense and Disarmament. The Committee will provide central review and coordination of activities in the Executive branch designed to improve our understanding of the economic impact of changes in defense expenditures. The President has also asked the Congress to establish a Presidential commission to study the impact of technological change on the economy and to recommend measures for assuring the full benefits of technology while minimizing any adverse effects.

Rapid advances in technology and sharp changes in the direction and location of economic activity pose both challenges and problems for the Nation. Your Committee has clearly been willing to take a completely fresh look at these matters. You may be sure that the Committee's analysis and recommendations will be given thoughtful consideration by all of those in the Executive branch who are concerned with these problems.

Sincerely,
Lee C. White
Assistant Special Counsel
to the President

Mr. W.H. Ferry
The Ad Hoc Committee on the
Triple Revolution

The Triple Revolution

This statement is written in the recognition that mankind is at a historic conjuncture which demands a fundamental reexamination of existing values and institutions. At this time three separate and mutually reinforcing revolutions are taking place:

The Cybernation Revolution: A new era of production has begun. Its principles of organization are as different from those of the industrial era as those of the industrial era were different from the agricultural. The cybernation revolution has been brought about by the combination of the computer and the automated self-regulating machine. This results in a system of almost unlimited productive capacity which requires progressively less human labor. Cybernation is already reorganizing the economic and social system to meet its own needs.

The Weaponry Revolution: New forms of weaponry have been developed which cannot win wars but which can obliterate civilization. We are recognizing only now that the great weapons have eliminated war as a method for resolving international conflicts. The ever-present threat of total destruction is tempered by the knowledge of the final futility of war. The need of a "warless world" is generally recognized, though achieving it will be a long and frustrating process.

The Human Rights Revolution: A universal demand for full human rights is now clearly evident. It continues to be demonstrated in the civil rights movement within the United States. But this is only the local manifestation of a worldwide movement toward the establishment of social and political regimes in which every individual will feel valued and none will feel rejected on account of his race.

We are particularly concerned in this statement with the first of these revolutionary phenomena. This is not because we underestimate the significance of the other two. On the contrary, we affirm that it is the simultaneous occurrence and interaction of all three developments which make evident the necessity for radical alterations in attitude and policy. The adoption of just policies for coping with cybernation and for extending rights to all Americans is indispensable to the creation of an atmosphere in the U.S. in which the supreme issue, peace, can be reasonably debated and resolved.

The Negro claims, as a matter of simple justice, his full share in America's economic and social life. He sees adequate employment opportunities as a chief means of attaining this goal: The March on Washington demanded freedom *and* jobs. The Negro's claim to a job is not being met. Negroes are the hardest-hit of the many groups being exiled from the economy by cybernation. Negro unemployment rates cannot be expected to drop substantially. Promises of jobs are a cruel and dangerous hoax on hundreds of thousands of Negroes and whites alike who are especially vulnerable to cybernation because of age or inadequate education.

The demand of the civil rights movement cannot be fulfilled within the present context of society. The Negro is trying to enter a social community and a tradition of work-and-income which are in the process of vanishing even for the hitherto privileged white worker. Jobs are disappearing under the impact of highly efficient, progressively less costly machines.

The U.S. operates on the thesis, set out in the Employment Act of 1964, that every person will be able to obtain a job if he wishes to do so and that this job will provide him with resources adequate to live and maintain a family decently. Thus job-holding is the general mechanism through which economic resources are distributed. Those without work have access only to a minimal income, hardly sufficient to provide the necessities of life, and enabling those receiving it to function as only "minimum consumers." As a result, the goods and services which are needed by these crippled consumers, and which they would buy if they could, are not produced. This in turn deprives other workers of jobs, thus reducing their incomes and consumption.

Present excessive levels of unemployment would be multiplied several times if military and space expenditures did not continue to absorb 10% of the gross national product (i.e., the total goods and services produced). Some 6 to 8 million people are employed as a direct result of purchases for space and military activities. At least an equal number hold their jobs as an indirect result of military or space expenditures. In recent years, the military and space budgets have absorbed a rising proportion of national production and formed a strong support for the economy.

However, these expenditures are coming in for more and more criticism, at least partially in recognition of the fact that nuclear weapons have eliminated war as an acceptable method for resolving international conflicts. Early in 1964 President Johnson ordered a curtailment of certain military expenditures. Defense Secretary McNamara is closing shipyards, airfields, and Army bases, and Congress is pressing the National Space Administration to economize. The future of these strong props to the economy is not as clear today as it was even a year ago.

How the Cybernation Revolution Shapes Up

Cybernation is manifesting the characteristics of a revolution in production. These include the development of radically different techniques and the subsequent appearance of novel principles of the organization of production; a basic reordering of man's relationship to his environment; and a dramatic increase in total available and potential energy.

The major difference between the agricultural, industrial and cybernation revolutions is the speed at which they developed. The agricultural revolution began several thousand years ago in the Middle East. Centuries passed in the shift from a subsistence base of hunting and food-gathering to settled agriculture.

In contrast, it has been less than 200 years since the emergence of the industrial revolution, and direct and accurate knowledge of the new productive techniques has reached most of mankind. This swift dissemination of information is generally held to be the main factor leading to widespread industrialization.

While the major aspects of the cybernation revolution are for the moment restricted to the U.S., its effects are observable almost at once throughout the industrial world and large parts of the non-industrial world. Observation is rapidly followed by analysis and criticism. The problems posed by the cybernation revolution are part of a new era in the history of all mankind but they are first being faced by the people of the U.S. The way Americans cope with cybernation will influence the course of this phenomenon everywhere. This country is the stage on which the machines-and-man drama will first be played for the world to witness.

The fundamental problem posed by the cybernation revolution in the U.S. is that it invalidates the general mechanism so far employed to undergird people's rights as consumers. Up to this time economic resources have been distributed on the basis of contributions to production, with machines and men competing for employment on somewhat equal terms. In the developing cybernated system, potentially unlimited output can be achieved by systems of machines which will require little cooperation from human beings. As machines take over production from men, they absorb an increasing proportion of resources while the men who are displaced become dependent on minimal and unrelated government measures—unemployment insurance, social security, welfare payments.

These measures are less and less able to disguise a historic paradox: That a substantial proportion of the population is subsisting on minimal incomes, often below the poverty line, at a time when sufficient productive potential is available to supply the needs of everyone in the U.S.

Industrial System Fails to Provide for Abolition of Poverty

The existence of this paradox is denied or ignored by conventional economic analysis. The general economic approach argues that potential demand, which if filled would raise the number of jobs and provide incomes to those holding them, is underestimated. Most contemporary economic analysis states that all of the available labor force and industrial capacity is required to meet the needs of consumers and industry and to provide adequate public services: Schools, parks, roads, homes, decent cities, and clean water and air. It is further argued that demand could be increased, by a variety of standard techniques, to any desired extent by providing money and machines to improve the conditions of the billions of impoverished people elsewhere in the world, who need food and shelter, clothes and machinery and everything else the industrial nations take for granted.

There is no question that cybernation does increase the potential for the provision of funds to neglected public sectors. Nor is there any question that cybernation would make possible the abolition of poverty at home and abroad. But the industrial system does not possess any adequate mechanisms to permit these potentials to become realities. The industrial system was designed to produce an ever-increasing quantity of goods as efficiently as possible, and it was assumed that the distribution of the power to purchase these goods would occur almost automatically. The continuance of the income-through-jobs link as the only major mechanism for distributing effective demand—for granting the right to consume—now acts as the main brake on the almost unlimited capacity of a cybernated productive system.

Recent administrations have proposed measures aimed at achieving a better distribution of resources, and at reducing unemployment and underemployment. A few of these proposals have been enacted. More often they have failed to secure congressional support. In every case, many members of Congress have criticized the proposed measures as departing from traditional principles for the allocation of resources and the encouragement of production. Abetted by budget-balancing economists and interest groups they have argued for the maintenance of an economic machine based on ideas of scarcity to deal with the facts of abundance produced by cybernation. This time-consuming criticism has slowed the workings of Congress and has thrown out of focus for that body the inter-related effects of the triple revolution.

An adequate distribution of the potential abundance of goods and services will be achieved only when it is understood that the major economic problem is not how to increase production but how to distribute the abundance that is the great potential of cybernation. There is an urgent need for a fundamental change in the mechanisms employed to insure consumer rights.

Facts and Figures of the Cybernation Revolution

No responsible observer would attempt to describe the exact pace or the full sweep of a phenomenon that is developing with the speed of cybernation. Some aspects of this revolution, however, are already clear:

- The rate of productivity increase has risen with the onset of cybernation.
- An industrial economic system postulated on scarcity has been unable to distribute the abundant goods and services produced by a cybernated system or potential in it.
- Surplus capacity and unemployment have thus co-existed at excessive levels over the last six years.
- The underlying cause of excessive unemployment is the fact that the capability of machines is rising more rapidly than the capacity of many human beings to keep pace.
- A permanent impoverished and jobless class is established in the midst of potential abundance.

Evidence for these statements follows:

1. The increased efficiency of machine systems is shown in the more rapid increase in productivity per man-hour since 1960, a year that marks the first visible upsurge of the cybernation revolution. In 1961, 1962 and 1963, productivity per man-hour rose at an average pace above 3.5%—a rate well above both the historical average and the postwar rate.

Companies are finding cybernation more and more attractive. Even at the present early stage of cybernation, costs have already been lowered to a point where the price of a durable machine may be as little as one-third of the current annual wage-cost of the worker it replaces. A more rapid rise in the rate of productivity increase per man-hour can be expected from now on.

2. In recent years it has proved to increase demand fast enough to bring about the full use of either men or plant capacities. The task of developing sufficient additional demand promises to become more difficult each year. A $30 billion annual increase in gross national product is now required to prevent unemployment rates from rising. An additional $40 to $60 billion increase would be required to bring unemployment rates down to an acceptable level.

3. The official rate of unemployment has remained at or above 5.5% during the Sixties. The unemployment rate for teenagers has been rising steadily and now stands around 15%. The unemployment rate for Negro teenagers stands about 30%. The unemployment rate for teenagers in minority ghettoes sometimes exceeds 50%. Unemployment rates for

Negroes are regularly more than twice those for whites, whatever their occupation, educational level, age or sex. The unemployment position for other racial minorities is similarly unfavorable. Unemployment rates in depressed areas often exceeds 50%.

Unemployment Is Far Worse Than Figures Indicate

These official figures seriously underestimate the true extent of unemployment. The statistics take no notice of underemployment or featherbedding. Besides the 5.5% of the labor force who are officially designated as unemployed, nearly 4% of the labor force sought full-time work in 1962 but could find only part-time jobs. In addition, methods of calculating unemployment rates—a person is counted as unemployed only if he has actively sought a job recently—ignore the fact that many men and women who would like to find jobs have not looked for them because they know there are no employment opportunities.

Underestimates for this reason are pervasive among groups whose unemployment rates are high—the young, the old, and racial minorities. Many people in the depressed agricultural, mining, and industrial areas, who by official definition hold jobs but who are actually grossly underemployed, would move if there were prospects of finding work elsewhere. It is reasonable to estimate that over 8,000,000 are not working who would like to have jobs today as compared with the 4,000,000 shown in the official statistics.

Even more serious is the fact that the number of people who have voluntarily removed themselves from the labor force is not constant but increases continuously. These people have decided to stop looking for employment and seem to have accepted the fact that they will never hold jobs again. This decision is largely irreversible, in economic and also in social and psychological terms. The older worker calls himself "retired"; he cannot accept work without affecting his social security status. The worker in his prime years is forced onto relief: In most states the requirements for becoming a relief recipient bring about such fundamental alterations in an individual's situation that a reversal of the process is always difficult and often totally infeasible. Teenagers, especially "drop-outs" and Negroes, are coming to realize that there is no place for them in the labor force but at the same time they are given no realistic alternative. These people and their dependents make up a large part of the "poverty" sector of the American population.

Statistical evidence of these trends appears in the decline in the proportion of people claiming to be in the labor force—the so-called labor force participation rate. The recent apparent stabilization of the unemployment rate around 5.5% is therefore misleading: It is a reflection of the discouragement and defeat of people who cannot find employment and have withdrawn from the market rather than a measure of the economy's success in creating jobs for those who want to work.

4. An efficiently functioning industrial system is assumed to provide the great majority of new jobs through the expansion of the private enterprise sector. But well over half of the new jobs created during 1957-1962 were in the public sector—predominantly in teaching. Job creation in the private sector has now almost entirely ceased except in services; of the 4,300,000 jobs created in this period, only about 200,000 were provided by private industry through its own efforts. Many authorities anticipate that the application of cybernation to certain service industries, which is only just beginning, will be particularly effective. If this is the case, no significant job creation will take place in the private sector in coming years.

5. Cybernation raises the level of the skills of the machine. Secretary of Labor Wirtz has recently stated that the machines being produced today have, on the average, skills equivalent to a high school diploma. If a human being is to compete with such machines, therefore, he must at least possess a high school diploma. The Department of Labor estimates, however, that on the basis of present trends, as many as 30% of all students will be high school drop-outs in this decade.

6. A permanently depressed class is developing in the U.S. Some 38,000,000 Americans, almost one-fifth of the nation, still live in poverty. The percentage of total income received by the poorest 20% of the population was 4.9% in 1944 and 4.7% in 1963.

Secretary Wirtz recently summarized these trends. "The confluence of surging population and driving technology is splitting the American labor force into tens of millions of 'have's' and millions of 'have-nots.' In our economy of 69,000,000 jobs, those with wanted skills enjoy opportunity and earning power. But the others face a new and stark problem—exclusion on a permanent basis, both as producers and consumers, from economic life. This division of people threatens to create a human slag heap. We cannot tolerate the development of a separate nation of the poor, the unskilled, the jobless, living within another nation of the well-off, the trained and the employed."

New Consensus Needed

The stubbornness and novelty of the situation that is conveyed by these statistics is now generally accepted. Ironically, it continues to be assumed that it is possible to devise measures which will reduce unemployment to a minimum and thus preserve the over-all viability of the present productive system. Some authorities have gone so far as to suggest that the pace of technological change should be slowed down "so as to allow the industrial productive system time to adapt."

We believe, on the contrary, that the industrial productive system is no longer viable. We assert that the only way to turn technological change to the benefit of the individual and the service of the general welfare is to accept the process and to utilize it rationally and humanely. The new science of political economy will be built on the encouragement and planned expansion of cybernation. The issues raised by cybernation are particularly amenable to intelligent policy-making: Cybernation itself provides the resources and tools that are needed to ensure minimum hardship during the transition process.

But major changes must be made in our attitudes and institutions in the foreseeable future. Today Americans are being swept along by three simultaneous revolutions while assuming they have them under control. In the absence of real understanding of any of these phenomena, especially of technology, we may be allowing an efficient and dehumanized community to emerge by default. Gaining control of our future requires the conscious formation of the society we wish to have. Cybernation at last forces us to answer the historic questions: What is man's role when he is not dependent upon his own activities for the material basis of his life? What should be the basis for distributing individual access to national resources? Are there other proper claims on goods and services besides a job?

Because of cybernation, society no longer needs to impose repetitive and meaningless (because unnecessary) toil upon the individual. Society can now set the citizen free to make his own choice of occupation and vocation from a wide range of activities not now fostered by our value system and our accepted modes of "work." But in the absence of such a new consensus about cybernation, the nation cannot begin to take advantage of all that it promises for human betterment.

Proposal for Action

As a first step to a new consensus it is essential to recognize that the traditional link between jobs and incomes is being broken. The economy of abundance can sustain all citizens in comfort and economic security whether or not they engage in what is commonly reckoned as work. Wealth produced by machines rather than by men is still wealth. We urge, therefore, that society, through its appropriate legal and governmental institutions, undertake an unqualified commitment to provide every individual and every family with an adequate income as a matter of right.

This undertaking we consider to be essential to the emerging economic, social and political order in this country. We regard it as the only policy by which the quarter of the nation now dispossessed and soon-to-be dispossessed by lack of employment can be brought within the abundant society. The unqualified right to an income would take the place of the patchwork of welfare measures—from unemployment insurance to relief—designed to ensure that no citizen or resident of the U.S. actually starves.

We do not pretend to visualize all of the consequences of this change in our values. It is clear, however, that the distribution of abundance in a cybernated society must be based on criteria strikingly different from those of an economic system based on scarcity. In retrospect, the establishment of the right to an income will prove to have been only the first step in the reconstruction of the value system of our society brought on by the triple revolution.

The present system encourages activities which can lead to private profit and neglects those activities which can enhance the wealth and the quality of life of our society. Consequently, national policy has hitherto been aimed far more at the welfare of the productive process than at the welfare of people. The era of cybernation can reverse this emphasis. With public policy and research concentrated on people rather than processes we believe that many creative activities and interests commonly thought of as non-economic will absorb the time and the commitment of many of those no longer needed to produce goods and services.

Society as a whole must encourage new modes of constructive, rewarding and ennobling activity. Principal among these are activities such as teaching and learning that relate people to people rather than people to things. Education has never been primarily conducted for profit in our society; it represents the first and most obvious activity inviting the expansion of the public sector to meet the needs of this period of transition.

We are not able to predict the long-run patterns of human activity and commitment in a nation when fewer and fewer people are involved in production of goods and services, nor are we able to forecast the overall patterns of income distribution that will replace those of the past full employment system. However, these are not speculative and fanciful matters to be contemplated at leisure for a society that may come into existence in three or four generations. The outlines of the future press sharply into the present. The problems of joblessness, inadequate incomes, and frustrated lives confront us now; the American Negro, in his rebellion, asserts the demands—and the rights—of all the disadvantaged. The Negro's is the most insistent voice today, but behind him stand the millions of impoverished who are beginning to understand that cybernation, properly understood and used, is the road out of want and toward a decent life.

The Transition*

We recognize that the drastic alternations in circumstances and in our way of life ushered in by cybernation and the economy of abundance will not be completed overnight. Left to the ordinary forces of the market such change, however, will involve physical and psychological misery and perhaps political chaos. Such misery is already clearly evident among the unemployed, among relief clients into the third generation and more and more among the young and the old for whom society appears to hold no promise of dignified or even stable lives. We must develop programs for this transition designed to give hope to the dispossessed and those cast out by the economic system, and to provide a basis for the rallying of people to bring about those changes in political and social institutions which are essential to the age of technology.

The program here suggested is not intended to be inclusive but rather to indicate its necessary scope. We propose:

1. A massive program to build up our educational system, designed especially with the needs of the chronically under-educated in mind. We estimate that tens of thousands of employment opportunities in such areas as teaching and research and development, particularly for younger people, may be thus created. Federal programs looking to the training of an additional 100,000 teachers annually are needed.

2. Massive public works. The need is to develop and put into effect programs of public works to construct dams, reservoirs, ports, water and air pollution facilities, community recreation facilities. We estimate that for each $1 billion per year spent on public works 150,000 to 200,000 jobs would be created. $2 billion or more a year should be spent in this way, preferably as matching funds aimed at the relief of economically distressed or dislocated areas.

3. A massive program of low-cost housing, to be built both publicly and privately, and aimed at a rate of 700,000-1,000,000 unites a year.

4. Development and financing of rapid transit systems, urban and interurban; and other programs to cope with the spreading problems of the great metropolitan centers.

---

*This view of the transitional period is not shared by all the signers. Robert Theobald and James Boggs hold that the two major principles of the transitional period will be (1) that machines rather than men will take up new conventional work openings and (2) that the activity of men will be directed to new forms of "work" and "leisure." Therefore, in their opinion, the specific proposals outlined in this section are more suitable for meeting the problems of the scarcity-economic system than for advancing through the period of transition into the period of abundance.

5. A public power system built on the abundance of coal in distressed areas, designed for low-cost power to heavy industrial and residential sections.

6. Rehabilitation of obsolete military bases for community or educational use.

7. A major revision of our tax structure aimed at redistributing income as well as apportioning the costs of the transition period equitably. To this end an expansion of the use of excess profits tax would be important. Subsidies and tax credit plans are required to ease the human suffering involved in the transition of many industries from man power to machine power.

8. The trade unions can play an important and significant role in this period in a number of ways:

a. Use of collective bargaining to negotiate not only for people at work but also for those thrown out of work by technological change.

b. Bargaining for perquisites such as housing, recreational facilities, and similar programs as they have negotiated health and welfare programs.

c. Obtaining a voice in the investment of the unions' huge pension and welfare funds, and insisting on investment policies which have as their major criteria the social use and function of the enterprise in which the investment is made.

d. Organization of the unemployed so that these voiceless people may once more be given a voice in their own economic destinies, and strengthening of the campaigns to organize white-collar and professional workers.

9. The use of the licensing power of government to regulate the speed and direction of cybernation to minimize hardship; and the use of minimum wage power as well as taxing powers to provide the incentives for moving as rapidly as possible toward the goals indicated by this paper.

These suggestions are in no way intended to be complete or definitively formulated. They contemplate expenditures of several billions more each year than are now being spent for socially rewarding enterprises, and a larger role for the government in the economy than it has now or has been given except in times of crisis. In our opinion, this is a time of crisis, the crisis of a triple revolution. Public philosophy for the transition must rest on the conviction that our economic, social and political institutions exist for the use of man and that man does not exist to maintain a particular economic system. This philosophy centers on an understanding that governments are instituted among men for the purpose of making possible life, liberty, and the pursuit of happiness and that government should be a creative and positive instrument toward these ends.

Change Must Be Managed

The historic discovery of the post-World War II years is that the economic destiny of the nation can be managed. Since the debate over the Employment Act of 1946 it has been increasingly understood that the federal government bears primary responsibility for the economic and social well-being of the country. The essence of management is planning. The democratic requirement is planning by public bodies for the general welfare. Planning by private bodies such as corporations for their own welfare does not automatically result in additions to the general welfare, as the impact of cybernation on jobs has already made clear.

The hardships imposed by sudden changes in technology have been acknowledged by Congress in proposals for dealing with the long and short-run "dislocations," in legislation for depressed and "impacted" areas, retraining of workers replaced by machines, and the like. The measures so far proposed have not been "transitional" in conception. Perhaps for this reason they have had little effect on the situations they were designed to alleviate. But the primary weakness of this legislation is not ineffectiveness but incoherence. In no way can these disconnected measures be seen as a plan for remedying deep ailments but only, so to speak, as the superficial treatment of surface wounds.

Planning agencies should constitute the network through which pass the stated needs of the people at every level of society, gradually building into a national inventory of human requirements, arrived at by democratic debate of elected representatives.

The primary tasks of the appropriate planning institutions should be:

- To collect the data necessary to appraise the effects, social and economic, of cybernation at different rates of innovation.
- To recommend ways, by public and private initiative, of encouraging and stimulating cybernation.
- To work toward optimal allocations of human and natural resources in meeting the requirements of society.
- To develop ways to smooth the transition from a society in which the norm is full employment within an economic system based on scarcity, to one in which the norm will be either non-employment, in the traditional sense of productive work, or employment on the great variety of socially valuable but "non-productive" tasks made possible by an economy of abundance; to bring about the conditions in which men and women no longer needed to produce goods and services

may find their way to a variety of self-fulfilling and socially useful occupations.

- To work out alternatives to defense and related spending that will commend themselves to citizens, entrepreneurs and workers as a more reasonable use of common resources.
- To integrate domestic and international planning. The technological revolution has related virtually every major domestic problem to a world problem. The vast inequities between the industrialized and the underdeveloped countries cannot long be sustained.

The aim throughout will be the conscious and rational direction of economic life by planning institutions under democratic control.

In this changed framework the new planning institutions will operate at every level of government—local, regional and federal—and will be organized to elicit democratic participation in all their proceedings. These bodies will be the means for giving direction and content to the growing demand for improvement in all departments of public life. The planning institutions will show the way to turn the growing protest against ugly cities, polluted air and water, an inadequate educational system, disappearing recreational and material resources, low levels of medical care, and the haphazard economic development into an integrated effort to raise the level of general welfare.

We are encouraged by the record of the planning institutions both of the Common Market and of several European nations and believe that this country can benefit from studying their weaknesses and strengths.

A principal result of planning will be to step up investment in the public sector. Greater investment in this area is advocated because it is overdue, because the needs in this sector comprise a substantial part of the content of the general welfare, and because they can be readily afforded by an abundant society. Given the knowledge that we are now in a period of transition it would be deceptive, in our opinion, to present such activities as likely to produce full employment. The efficiencies of cybernation should be as much sought in the public as in the private sector, and a chief focus of planning would be one means of bringing this about. A central assumption of planning institutions would be the central assumption of this statement, that the nation is moving into a society in which production of goods and services is not the only or perhaps the chief means of distributing income.

The Democratization of Change

The revolution in weaponry gives some dim promise that mankind may finally eliminate institutionalized force as the method of settling international conflict and find for it political and moral equivalents leading to a better world. The Negro revolution signals the ultimate admission of this group to the American community on equal social, political and economic terms. The cybernation revolution proffers an existence qualitatively richer in democratic as well as material values. A social order in which men make the decisions that shape their lives becomes more possible now than ever before; the unshackling of men from the bonds of unfulfilling labor frees them to become citizens, to make themselves and to make their own history.

But these enhanced promises by no means constitute a guarantee. Illuminating and making more possible the "democratic vistas" is one thing; reaching them is quite another, for a vision of democratic life is made real not by technological change but by men consciously moving toward that ideal and creating institutions that will realize and nourish the vision in living form.

Democracy, as we use the term, means a community of men and women who are able to understand, express and determine their lives as dignified human beings. Democracy can only be rooted in a political and economic order in which wealth is distributed by and for people, and used for the widest social benefit. With the emergence of the era of abundance we have the economic base for a true democracy of participation, in which men no longer need to feel themselves prisoners of social forces and decisions beyond their control or comprehension.

Signers

Todd Gitlin
President, Students for
a Democratic Society

Roger Hagan Editor,
The Correspondent,
Council for Correspondence

Michael Harrington
Author, *The Other America*

Tom Hayden
Students for a Democratic Society

Ralph L. Helstein
President, United Packinghouse,
Food and Allied Workers

Dr. Frances W. Herring
Institute for Governmental Studies
University of California

Brig. Gen. Hugh B. Hester
Director of Procurement of
Supplies for Gen. MacArthur's forces in the SW Pacific, 1942-47

Gerald W. Johnson
Journalist

Irving F. Laucks
Former head,
Laucks Chemical Company,
Consultant to The Center for the
Study of Democratic Institutions

Gunner Myrdal
Economist,
Institute for International Economic Studies
"I am in broad agreement with this Statement, though not entirely so."

Gerard Piel
Publisher,
Scientific American

Michael D. Reagan
Graduate Program in Public Administration, Maxwell Graduate School of Citizenship and Public Affairs, Syracuse University

Ben B. Seligman
Director, Department of Education and Research, Retail Clerks International Association

Robert Theobald
Consulting economist, and author of many books, the latest being *Free Men and Free Markets*

William Worthy
Correspondent,
Baltimore *Afro-American*

Alice Mary Hilton
Independent consultant,
problems of technology
and automation

Maxwell Geismar
Social critic,
author of *Henry James and the Jacobites*

Philip Green
Assistant Professor of Political Science,
Haverford College

H. Stuart Hughes
Professor of History,
Harvard University

Linus Pauling
Nobel laureate, Peace;
Consultant to the Center for the Study of Democratic Institutions

John William Ward
Professor of History,
Princeton University

A. J. Muste
Secretary Emeritus,
Fellowship of Reconciliation

Dr. Louis Fein
Independent consultant,
computer field

Stewart Meacham
Peace Secretary, American Friends Service Committee

Everett C. Hughes
Professor of Sociology,
Brandeis University

Robert L. Heilbroner
Economist; author, *The Worldly Philosophers, The Great Ascent*

Irving Howe
Editor, *Dissent* Magazine

Bayard Rustin
Executive Secretary, War Resisters League; organizer of the March on Washington

Norman Thomas
Socialist leader

Dwight Macdonald
Writer and critic

Carl F. Stover
Director,
National Institute of Public Affairs

Donald G. Agger
Attorney, de Grazia, Agger and Hydeman

Dr. Donald B. Armstrong, M.D.

James Boggs
author of *Pages from a Negro Workers's Notebook*

W.H. Ferry
Vice-President, Fund for the Republic

Appendix B

Participants in the Conference on "Technology for the Common Good" 4-5 October 1991

Gar Alperovitz
Fellow, Institute for Policy Studies

Richard Barnet
Distinguished Fellow,
Institute for Policy Studies

Gary Bass
Executive Director, OMB Watch

Eddie Becker
Independent Television Producer

Scott Bernstein
Executive Director, Center for Neighborhood Technology

Greg Bischak
Executive Director, National Commission for Economic Conversion and Disarmament

Robert Borosage
Fellow, Institute for Policy Studies

John Cavanagh
Fellow, Institute for Policy Studies

Gary Chapman
Executive Director, Computer Professionals for Social Responsibility

Dan Charles
New Scientist

Richard Civille
Computer Professionals for Social Responsibility

Gary Cohen
Co-Director, National Toxics Campaign

W.H. Ferry
Social Critic

Wade Greene
Executive Director,
Rockefeller Family Associates

Chellis Glendinning
Author, *When Technology Wounds*

Sandra Hackman
Managing Editor, *Technology Review*

Hal Harvey
Executive Director,
Energy Foundation

Richard Healey
Fellow, Institute for Policy Studies

David H. Horowitz
Proskauer, Rose,
Goetz and Mendelsohn

Bill Keepin
Energy Policy Consultant

Saul Landau
Fellow, Institute for Policy Studies

Vincent McGee
Executive Director,
Aaron Diamond Foundation

Gisele Mills
Fellow, Institute for Policy Studies

Kathryn Montgomery & Jeff Chester
Directors, Center for
Media Education

Ralph Nader
Director, Center for Study
of Responsive Law

Doreen Nelson
Center for City Building Educational Program

Marc Raskin
Distinguished Fellow,
Institute for Policy Studies

Charley Richardson
Director, Technology and Work
Program, University of
Massachusetts, Lowell

Marc Rotenberg
Washington Office Director, Computer
Professionals for Social Responsibility

Rustum Roy
Professor of Solid State Engineering at
Penn State

Ray Scannell
Director of Research, Bakery,
Confectionary, and Tobacco Workers
International

Richard Sclove
Executive Director, Loka Institute

Michael Shuman
Executive Director,
Institute for Policy Studies

William Worthy
Professor of Communications,
Howard University

Jeffrey Young
The Nation

Joel Yudken
Specialist, Technology Policy and Economic Conversion

Contributors

Chellis Glendinning is a psychologist whose focus is the interplay between personal psychology and social issues. She is Adjunct Professor at the California Institute of Integral Studies and the author of the Pulitzer Prize-nominated *When Technology Wounds* (Morrow). She also wrote *Waking Up in the Nuclear Age* (New Society). Glendinning lives in Tesuque, New Mexico, and is currently working on a book about addiction and the ecological crisis called *My Name is Chellis. I'm In Recovery From Western Civilization.*

Sanford Lewis is the director of the Good Neighbor Project of the Center for the Study of Public Policy, an effort to build community-corporate partnerships for pollution prevention and job security. He is also an instructor of environmental law at Tufts University and has served as the director of technical and legal support for the National Toxics Campaign Fund. The author wishes to thank Birgit Caliandro, Susanne Rasmussen and Heidi Porter for their assistance in preparing this paper.

Charley Richardson is the director of the Technology and Work Program at the University of Massachusetts, Lowell. The program provides training, technical assistance, and strategic-planning support to unions and workers dealing with the impact of technological change. Richardson worked for ten years as a heavy steel fabricator in three diffferent shipyards. He was a steward in the boilermakers union and a safety observer for the International Union of Marine and Shipbuilding Workers of America (IUMSWA) Local 5, at General Dynamics in Quincy, Massachusetts.

Richard Sclove is executive director of the Amherst, Massachusetts-based Loka Institute, an association of scholars and activists concerned with the social and political ramifications of science, technology, and architecture. The author of a forthcoming book, *Technology and Freedom,* he is also a contributor to the recent anthologies, *Democracy in a Technological Society* (Kluwer Academic Publishers, 1992) and *Critical Perspectives on Non-Academic Science and Engineering* (Lehigh University Press, 1991). He holds advanced degrees in nuclear engineering and political science. Dr. Sclove is currently an advisor to the Telecommunications & Democracy Project at Rutgers University; Computer Professionals for Social Responsibility's 21st

Century Project, and the Science, Technology & Society Program at the University of Massachusetts-Amherst.

Michael Shuman, a 36 year old lawyer, is executive director of the Institute for Policy Studies in Washington, D.C., and a specialist in citizen and city participation in international affairs. Shuman has co-written one book, *Citizen Diplomats: Pathfinders in Soviet-American Relations* (Continuum, 1986), co-edited a second, *Conditions of Peace: An Inquiry* (Expro Press, 1992), and is currently completing two others — *Security without War: A Post-Cold War Foreign Policy* (Westview, 1993) and *Local Rights and Global Wrongs: The Legality of Municipal Foreign Policy*. This last book was supported by a research and writing fellowship from the John D. and Catherine T. MacArthur Foundation. He has also written articles for such periodicals as *Bulletin of the Atomic Scientists, Foreign Policy, Parade, New York Times*, and *Foundation News*. Between 1987 and 1990 he was a W.K. Kellogg National Leadership Fellow.

Julia Sweig is a doctoral candidate at the Johns Hopkins University School of Advanced International Studies (SAIS) and a research fellow at the Institute for Policy Studies. She is currently the editor of *CubaINFO*, the publication of record on Cuba and U.S.-Cuban relations, published at Johns Hopkins University. She is co-editor of *Conditions of Peace: An Inquiry*, (Expro Press, 1992).

Joel Yudken is currently working in the United States Senate as a Congressional Science and Engineering Fellow sponsored by the American Association for the Advancement of Science. From 1990 to 1992, he was a research fellow at the Project on Regional and Industrial Economics at Rutgers University, in New Brunswick, New Jersey. An electronics engineer by training, who has worked in the defense industry, Yudken holds a Masters degree in engineering-economic systems and a Ph. D. in technology and society from Stanford University. He is co-author, with Ann Markusen, of *Dismantling the Cold War Economy* (Basic Books, 1992), and has published numerous articles on technology and industrial policy, and economic conversion.

The Institute for Policy Studies

The Institute for Policy Studies (IPS) is the nation's leading think thank on the progressive side of the political spectrum. It promotes public policies and grassroots action consistent with the goals of economic justice, social equity, ecological sustainability, democratic participation, personal freedom, human dignity, world peace, community empowerment, and global responsibility.

Since being established in 1963 by Marcus Raskin and Richard Barnet, IPS has helped develop the intellectual foundations for numerous citizens movements, including those opposing U.S. military intervention in Vietnam and Central America, promoting civil rights and women's rights, challenging the axioms of the "national security state" and developing alternatives to free trade. Its public scholars have consistently provided informed critiques of conventional wisdom and presented constructive alternatives through books, articles, op-eds, speeches, films, congressional testimony, and radio and television appearances.

Even though the public scholars at IPS share many goals, they do not subscribe to any one school of thought or "ism." Indeed, over the years, IPS has been a home for a wide range of gifted public scholars, including liberals, social democrats, socialists, communitarians, feminists, libertarians, greens, and anarchists.

The bylaws of the Institute state that its public scholars shall have complete intellectual freedom and, consequently, that none of their views—including those expressed in this book—shall be attributed to the institution as a whole or to its board of trustees. To further ensure their independence, IPS fellows do not solicit or accept grants or any other remunerations form the national government. The work of IPS is supported solely through the contributions of its members and private foundations.

If you are interested in becoming a member of the Institute, please send us your name, address, and phone number along with a check for $25 to cover annual dues. Members receive the IPS quarterly newsletter and discounts on all IPS publications. For more information, contact IPS at:

Institute for Policy Studies,
1601 Connecticut Ave., NW Fifth Floor
Washington, DC 20009
(202) 234-9382 (Office)
202-387-7915 (Fax)

Other Publications from IPS

Plundering Paradise
The Struggle for the Environment in the Philippines
by Robin Broad & John Cavanagh
University of California Press,
1993, 197 pp., cloth
ISBN 0-520-08081-5, $25.00

Abolishing the War System
Marcus G. Raskin, Project Director
The Disarmament and International Law Project of the Institute for Policy Studies and the Lawyers Committee on Nuclear Policy
Aletheia Press,
1993, 129 pp., pbk
ISBN 0-9623718-8-2, $12.00

Trading Freedom
How Free Trade Affects our Lives, Work and Environment
John Cavanagh, John Gershman, Karen Baker & Gretchen Helmke eds.
IPS & IFDP,
1992, 138 pp., pbk
ISBN 0-935028-59-5, $10.00

Conditions of Peace
An Inquiry
Michael Shuman & Julia Sweig, eds.
EXPRO,
1991, 254 pp., pbk
ISBN 0-93639140-5,
Originally $15.95
Reduced to $10.00

The Guerrilla Wars of Central America
Nicaragua, El Salvador & Guatemala
by Saul Laundau
Widenfeld & Nicholson,
1993, 211 pp. cloth
ISBN 0-29782-114-8, $35.00

Who Pays? Who Profits?
The Truth about the American Tax System
by Ralph Estes
IPS,
1993, 125 pp., pbk
ISBN 0-89758-048-6, $5.95

Paradigms Lost
The Post Cold War Era
Chester Hartman & Pedro Vilanova, eds.
IPS & TNI,
1992, 205 pp., pbk.
ISBN 0-7453-0638-1, $12.00

The Debt Boomergang
How third World Debt Harms Us All
by Susan George
Pluto Press,
1992, 202 pp., pbk
ISBN 0-7454-0594-6, $6.00

Order Form

Ship to:

Name ____

Address ____

____

City ____________________ State __________ Zip __________

Phone (   ) _______________________________________

Qty	Title	Price
	Subtotal	
	Shipping	
	Donation	
	Total	

Shipping Costs

For orders under $30.00, add 20%; $30-$100, add 10%; over $100, add 5%.

Address orders and checks to:
IPS Publications
1601 Connecticut Ave., NW
Washington, DC 20009

☐ Please send me more information about the Institute for Policy Studies.

For convenient credit card orders call (202) 234-9382 ext. 206